Proof Patterns

Mark Joshi

Proof Patterns

 Springer

Mark Joshi
Centre for Actuarial Studies
University of Melbourne
Melbourne, VIC
Australia

ISBN 978-3-319-16249-2   ISBN 978-3-319-16250-8   (eBook)
DOI 10.1007/978-3-319-16250-8

Library of Congress Control Number: 2015932521

Mathematics Subject Classification: 00–01, 97D50

Springer Cham Heidelberg New York Dordrecht London

Printed on acid-free paper

Springer International Publishing AG Switzerland is part of Springer Science+Business Media
(www.springer.com)

Preface

Patterns have become a common theme in many fields of academic study. In programming, in particular, the book "Design Patterns" has become highly influential and it is now customary to discuss programs in terms of the patterns used. The programmers generally attribute the idea of a design pattern to architecture. The fundamental idea is that each field has a collection of ways of breaking down problems into component pieces. Understanding these methodologies explicitly then leads to greater comprehension, facilitates learning and simplifies problem solving. Rather than attempting a problem cold, one first sees whether known patterns work. Even if they all fail, understanding why they fail defines the problem.

In this book, my objective is to identify and teach many of the common patterns that arise in pure mathematics. I call these "Proof Patterns". The main originality in the presentation is that examples are focussed about each pattern and drawn from different areas. This differs from the usual style of teaching pure mathematics where a topic is chosen and dissected; patterns are then drawn in as needed, and they are often not explicitly mentioned. After studying enough topics the learner picks up a variety of patterns, and the difficulty of studying a new area is often determined by the degree of unfamiliarity with its patterns.

This book is intended to do a variety of things. On one level, my objective is to teach the basic patterns. On another, it is intended as a taster for pure mathematics. The reader will gain a little knowledge on a variety of topics and hopefully learn a little about what pure mathematics is. On a third level, the intention of the book is to make a case for the explicit recognition of patterns when teaching pure mathematics. On a fourth level, it is simply an enjoyable romp through topics I love.

One powerful tool of pure mathematics which I intentionally avoid is that of abstraction. I believe that patterns and concepts are best learnt via the study of concrete objects wherever possible. Whilst one must go abstract eventually to obtain the full power and generality of results, a proof or pattern that has already been understood in a concrete setting is much easier to comprehend and apply.

A side effect of this avoidance is that the patterns are not formally defined since such definitions would require a great deal of abstraction.

The target reader of this book will already be familiar with the concept of proof but need not know much more. So whilst I assume very few results from pure mathematics, the reader who does not know what a proof is will struggle. There are several excellent texts such as Eccles, Velleman and Houston for such readers to study before reading here. In particular, I regard this book as a second book on proof, and my hope is that the reader will find that the approach here eases their study of many areas of pure mathematics.

I try to build up everything from the ground up as much as possible. I therefore try to avoid the "pull a big theorem out of the hat" style of mathematics presentation. The emphasis is much more on how to prove results rather than on trying to impress with theorems whose proofs are far beyond the book's scope. I do occasionally use concepts from analysis before these are formally defined such as a convergent sequence. Hopefully, the reader who has not studied analysis will be able to work with their intuitive notions of these objects.

Inevitably, as with many introductory books on proof, many examples are drawn from combinatorics and elementary number theory. This reflects the fact that these areas require fewer prerequisites than most and so patterns can be discussed in simple settings. However, I also draw on a variety of areas including group theory, linear algebra, computer science, analysis, topology, Euclidean geometry, and set theory to emphasize patterns' universality.

There is little if any originality in the mathematical results in this book: the objective was to provide a different presentation rather than new results. We look at the "Four-Colour problem" at various points. Our treatment is very much inspired by Robin Wilson's excellent book "Four colours suffice" and I recommend it to any reader whose interest has been piqued. A book with some similarities to this one but requiring a little more knowledge from the reader is "Proofs from the BOOK" by Aigner and Ziegler. The emphasis there is more on beauty in proof than on patterns and it is a good follow on for the reader who wants more. However, I do hope that any reader of this book will develop some appreciation for the beauty of mathematics.

Many of the patterns in this book have not been named before although they are in widespread use. I have therefore invented their names. I hope that these new names will prove popular. I apologise to those who dislike them.

For the reader who has forgotten or never knew mathematical terminology, I have included a glossary in Appendix A. This also includes definitions of the standard sets of numbers. For clarity, let me say right here that in this book 0 is a natural number. This is the way I was taught as an undergraduate and it is too firmly embedded in my psyche for me to use any other definition. The term *counting numbers* denoted $\mathbb{N}_1$ will be used for the natural numbers excluding zero.

This book is ultimately an expression of my philosophy of how to approach the teaching of mathematics. My views have been shaped by interactions with innumerable former teachers, students and colleagues and I thank them all. I particularly thank Alan Beardon and Navin Ranasinghe for their detailed comments on a former version of the text. I also thank some anonymous referees for their constructive comments.

Melbourne, 2014   Mark Joshi

Contents

Chapter 1
Induction and Complete Induction

1.1 Introduction

Two basic proof tools are the principle of induction and the related principle of complete induction. Each of these relies on certain properties of the natural numbers which we now explore.

The idea of induction is simple. We wish to prove that a set of statements, $P(n)$, hold for all $n \in \mathbb{N}$ greater than or equal to some natural number k. We first check it for $P(k)$, and then we show that if it holds for one value of n then it also holds for the next one. In other words, we have to prove

- $P(k)$;
- $P(n) \implies P(n+1)$, $\forall n \geqslant k$.

It then follows that it holds for all values of n starting with k. Why? It holds for k so putting $n = k$, it holds for $k+1$. Putting $n = k+1$, it holds for $k+2$. Repeating, it holds for $k+3, k+4, k+5, \ldots$. We discuss how to show that $P(n)$ really does hold for $n \geqslant k$ in Sect. 1.3.

Complete induction is a closely related tool. The difference is that we are allowed to use $P(l)$ for $k \leqslant l \leqslant n$ when proving $P(n+1)$ rather than just $P(n)$. This can be advantageous—for certain types of results, it is the truth of $P(l)$ for some much smaller l that is useful rather than that of $P(n)$. It holds for similar reasons to ordinary induction and we will discuss the proofs that they hold together.

© Springer International Publishing Switzerland 2015
M. Joshi, *Proof Patterns*, DOI 10.1007/978-3-319-16250-8_1

1.2 Examples of Induction

Induction often come in useful when establishing formulas for sums. Let $f(n)$ be the sum of the first n natural numbers. We want to prove

$$f(n) = \frac{n(n+1)}{2}.$$

In this case, $P(n)$ is the statement that the formula is correct for $f(n)$.

We certainly have $f(1) = 1$ so $P(1)$ is true. We now assume that $P(n)$ is true and try to establish that $P(n+1)$ is true. In this context $P(n)$ is sometimes called the *inductive hypothesis*. We have

$$f(n+1) = f(n) + n + 1,$$

by definition. We now substitute the formula which we *assumed* to be true for n, to obtain

$$f(n+1) = \frac{n(n+1)}{2} + (n+1) = \frac{n(n+1) + 2(n+1)}{2}.$$

Simplifying,

$$f(n+1) = \frac{(n+2)(n+1)}{2}.$$

We have shown that $P(n)$ implies $P(n+1)$ and the results holds for all n by induction.

Whilst proof by induction is often easy and in a case like this it will generally work if the result is true, it has the disadvantage that you have to already know the formula! It is therefore more a way of proving formulas rather than of finding them. It can therefore come in useful when a result has been guessed in some non-rigorous fashion and still has to be proven.

We now give a classic example of applying the principle of complete induction. We show that any natural number bigger than one can be written as a product of prime numbers. In this case, $P(n)$ is the statement that n can be written as a product of product of primes.

We start with $P(2)$. It certainly holds since 2 is prime. We now assume the inductive hypothesis that $P(2), P(3), \ldots, P(n)$ are true. Consider $n + 1$. Either it is prime in which case we are done, or it is composite. In the latter case, we can write

$$n + 1 = ab$$

with $2 \leqslant a, b \leqslant (n+1)/2 \leqslant n$. We have assumed that $P(a)$ and $P(b)$ hold so we can write

$$a = p_1 p_2 \ldots p_\alpha, \quad b = q_1 q_2 \ldots q_\beta$$

for some primes p_j and q_j and $\alpha, \beta \in \mathbb{N}$. So

$$n + 1 = p_1 p_2 \cdots p_\alpha q_1 q_2 \cdots q_\beta$$

and $P(n + 1)$ holds. It follows from complete induction, that all natural numbers bigger than 1 can be written as a product of primes. Note that this proof used complete induction in a non-trivial way; it is not at all clear how we could prove the result using only ordinary induction.

More generally, note that whilst we have established that every natural number bigger than 1 is a product of primes, we have not shown that the representation is unique. Indeed without some extra restrictions, it is not:

$$2 \times 2 \times 3 = 12 = 2 \times 3 \times 2.$$

This non-uniqueness can be circumvented by requiring the primes to be in ascending order, and the representation is then unique. Note the general technique here: impose additional structure to remove non-uniqueness. However, one still has to prove that this amount of extra structure is enough to gain uniqueness. We will prove that it is sufficient in Sect. 4.6.

1.3 Why Does Induction Hold?

How can we prove that induction works? One solution is to use the well-ordering of the natural numbers: every non-empty subset of the natural numbers has a smallest element. The idea is essentially that there cannot be a least element that the statement does not hold for, so the set of such elements must be empty.

More formally, the argument is let E be the set of n for $n > k$ for which $P(n)$ is false. If E is non-empty then it has a least element l and we know $l > k$. So $l - 1 \notin E$ so $P(l - 1)$ holds. By the inductive hypothesis, $P(l)$ holds. So l is both in E and not in E. We have a contradiction. So E has no least element and must be empty.

We can prove the principle of complete induction in a similar fashion. In fact, we can deduce it from the principle of induction directly. Let $Q(n)$ be the statement that $P(l)$ holds for $k \leqslant l \leqslant n$. We then have that $Q(k)$ holds, and that $Q(n)$ implies $Q(n + 1)$ so it follows by induction that $Q(n)$ holds for all $n \geqslant k$. Since $Q(n)$ certainly implies $P(n)$, we also have that $P(n)$ holds and we are done.

These arguments are correct; however, we have proven the principle of induction by assuming the well ordering of the natural numbers. How can we prove that well ordering holds? (Un)fortunately, proofs really come down to the fact that the principle of induction holds! We have to take one of the two as an axiom and use it to deduce the other. Ultimately, mathematicians define their formal systems and axioms in such a way as to capture their intuitive notions of what they are trying to model. In this case, the axioms capture the notion that all natural numbers can be reached by starting at 0 and repeatedly adding one.

1.4 Induction and Binomials

The *binomial coefficient* $\binom{n}{k}$ expresses the number of ways that k objects can be selected from n objects with $n \geqslant k$. We do not care about the order of the objects selected. We have

$$\binom{n}{k} = \frac{n!}{(n-k)!\,k!}.$$

Why? The number of ways we can select the first object is n. The second is then $n-1$ since one is gone, and $n-2$ for the one after, and so on. (See if you can write out a formal proof of this using induction.) So we can select k objects in

$$n(n-1)\ldots(n-k+1) = \frac{n!}{(n-k)!}$$

different ways. However, this is an ordered selection. We do not care about the ordering so we can divide again by the number of different orderings of k objects and we get

$$\frac{n!}{(n-k)!\,k!},$$

as desired. Note that an immediate consequence of our interpretation of this fraction is that it represents a whole number! We always have that

$$k! \mid n(n-1)\ldots(n-k+1)$$

which is not obvious. For the reader not familiar with $\mid$, we say that for $a, b \in \mathbb{Z}$

$$a \mid b$$

if there exists $m \in \mathbb{Z}$ such that
$$ma = b.$$

In other words, b/a is an integer.

Note that here, we have shown a relationship between two numbers by interpreting a formula in a certain way. Sometimes this can yield non-obvious properties: *proof by observation.*

We take $0! = 1$. This can be regarded as a definition, however, it makes sense in that $k!$ expresses the number of ways you can order k objects. If we have no objects then there is only one way to order them so we should have $0! = 1$. We have

$$\binom{n}{0} = 1 = \binom{n}{n};$$

we also have

$$\binom{n}{1} = n = \binom{n}{n-1}.$$

An obvious symmetry exists

$$\binom{n}{k} = \binom{n}{n-k}.$$

We can also show *Pascal's identity*

$$\binom{n}{k} + \binom{n}{k-1} = \binom{n+1}{k},$$

for $1 \leqslant k \leqslant n$. This can be proven either via algebraic manipulation or by interpretation. We use the latter approach. We want to show that the number of subsets with k elements taken from $\{1, 2, \ldots, n+1\}$ is the left-hand-side of the identity. We show that any such subset corresponds to either a subset with k elements of $1, 2, \ldots, n$ or one with $k-1$ elements. If our subset, E, with k elements contains the element $n+1$ then discarding $n+1$ gives a subset of $\{1, \ldots, n\}$ with $k-1$ elements. Clearly all such subsets can be obtained this way. If E does not contain $n+1$ then it is a subset of $\{1, 2, \ldots, n\}$ with k elements, and again we can get all such subsets in this way. We have constructed a correspondence and the identity follows.

With Pascal's identity in hand, we can now prove something using induction.

Theorem 1.1 *The binomial theorem. If n is a natural number, and x, y are real numbers then*

$$(x + y)^n = \sum_{k=0}^{n} \binom{n}{k} x^k y^{n-k}.$$

Proof If $n = 0$, both sides are equal to 1. Now suppose the result holds for n. We write

$$(x + y)^{n+1} = (x + y)(x + y)^n = x(x + y)^n + y(x + y)^n.$$

Using the inductive hypothesis,

$$(x + y)^{n+1} = \sum_{k=0}^{n} \binom{n}{k} \left(x^{k+1} y^{n-k} + x^k y^{n-k+1} \right).$$

We need to gather terms with the same powers of x and y together,

$$\binom{n}{k} x^{k+1} y^{n-k} = \binom{n}{l-1} x^l y^{n+1-l}$$

where $l = k + 1$. So, letting $\binom{n}{-1} = 0$,

$$(x + y)^{n+1} = \sum_{l=0}^{n} \left(\binom{n}{l} + \binom{n}{l-1} \right) x^l y^{n-l+1} + \binom{n}{n} x^{n+1}.$$

Invoking Pascal's identity and the fact that $\binom{n}{n} = 1 = \binom{n+1}{n+1}$, we have

$$(x + y)^{n+1} = \sum_{l=0}^{n+1} \binom{n+1}{l} x^l y^{n+1-l},$$

as required and the result follows by induction.   □

Induction is not essential for the proof of the binomial theorem. Another approach is to think about how the coefficient of $x^k y^{n-k}$ arises when we work out the expansion of $(x + y)^n$. It occurs from picking the x of $(x + y)$ in k places out of n possible ones. It can therefore arise $\binom{n}{k}$ different times and we get the binomial theorem.

With the binomial theorem proven, we can make various proofs by observation. First,

$$2^n = \sum_{k=0}^{n} \binom{n}{k}. \tag{1.4.1}$$

To prove this, just put $x = y = 1$ in the binomial theorem. Once we interpret this result, it is actually clear for other reasons. We are adding the number of ways of choosing k elements from n for each k. We are therefore counting the number of subsets of $\{1, 2, \ldots, n\}$. Each number is either in a given subset or not, so each number gives us two possibilities. There are n numbers so there are 2^n subsets and we have (1.4.1). Our alternate proof is a proof by *double counting*; we counted a collection of objects in two different ways to establish a formula.

Another immediate consequence of the binomial theorem is

$$\sum_{k=0}^{n} \binom{n}{k} (-1)^k = 0. \tag{1.4.2}$$

Just set $x = -1$ and $y = 1$. We can rewrite this as

$$\sum_{k=0}^{k \leqslant n/2} \binom{n}{2k} = \sum_{k=0}^{k < n/2} \binom{n}{2k+1}. \tag{1.4.3}$$

The number of different ways of choosing a subset with an even number of elements equals the number of ways of choosing a subset with an odd number. This is clear when n is odd; taking the complement provides a natural bijection between the two classes of subsets. It is not so obvious when n is even.

1.5 Triangulating Polygons

We now look at an application of complete induction to a quite different area. We prove that every polygon in the plane can be triangulated. In fact, we prove a stronger result, we show that it can be done without adding any extra vertices. Before proceeding to the proof, we discuss what a triangulation is. A polygon is a closed loop in the plane consisting of a sequence of straight line segments which starts and finishes at the same point. The loop does not self intersect anywhere. To triangulate means to divide the polygon into triangles which only intersect along common sides. We give a couple of examples in Fig. 1.1. We prove

Theorem 1.2 *If P is a polygon in the plane, it is possible to write P as a union of triangles whose vertices are subsets of those of P and which only intersect in common sides or vertices.*

Proof A polygon has at least 3 vertices. If it has 3 exactly then it is a triangle and the result is trivial. We now assume that any polygon with $3 \leqslant k < n$ sides can be triangulated. Let P be a polygon with n sides. It must have a vertex, V, where the interior angle between its two edges is less than $180°$. To see this observe that if the change in direction of the edge at a vertex is x degrees then the interior angle is $180 - x$ degrees. Since all the changes of direction must add up to $360°$, at least one must have $x > 0$ and so that angle must be less than 180.

Now consider the two vertices next to V. Call them A and B. We draw a line between them. See Fig. 1.2. If this line's interior does not intersect P then ABC defines a triangle that we can cut off P. The remaining part of P can be triangulated by the inductive hypothesis and so P is triangulable.

If the line AB's interior does intersect P, then we slide its endpoints along towards V. We do the sliding in such a way that they both reach V at the same time. Since

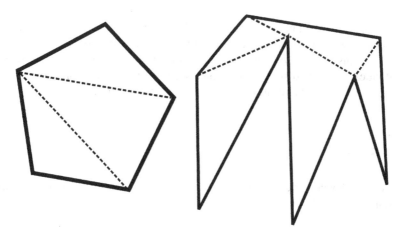

Fig. 1.1 Examples of triangulations of polygons

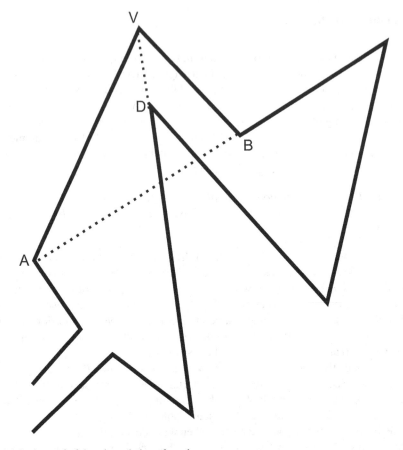

Fig. 1.2 A proof of the triangulation of a polygon

there are only a finite number of vertices in P, there will be a last time at which the line crosses a vertex. Call a vertex crossed at this last time D. (There could be more than one but this is not important.) The line from A to D will now not intersect P and it divides P into two smaller polygons. These can both be triangulated by the inductive hypothesis. The result now follows by complete induction.   □

1.6 Problems

Exercise 1.1 Develop an expression for the sum of the first n odd numbers and prove that it holds using induction.

Exercise 1.2 Prove that for $n \geqslant 1$,

$$\sum_{j=1}^{n} \frac{1}{\sqrt{j}} \geqslant \sqrt{n}.$$

Exercise 1.3 A chess-style board is of size $2^n \times 2^n$ and has a single square deleted. A trimino is three squares joined together so as to have an angle; it is like a domino with a square stuck to the side. Show that for any n, the board can be covered by non-overlapping triminoes.

Exercise 1.4 Prove that

$$\sum_{j=1}^{n} j^2 = \frac{n(n+1)(2n+1)}{6}.$$

Exercise 1.5 Prove that

$$\sum_{j=1}^{n} 2^{j-1} = 2^n - 1.$$

Exercise 1.6 Prove that if $x > -1$ then

$$(1+x)^n \geqslant 1 + nx,$$

for $n \in \mathbb{N}$.

Exercise 1.7 Prove that every number bigger than 11 is a positive integer combination of 4 and 5. That is if $k \in \mathbb{N}$, $k \geqslant 12$, there exists $a, b \in \mathbb{N}$ such that

$$k = 4a + 5b.$$

Exercise 1.8 A polynomial is reducible if it can be written as a product of polynomials which are not constant. If it is not reducible, it is said to be irreducible. Prove that every non-constant polynomial is a product of irreducible polynomials.

Exercise 1.9 Show that if $n \in \mathbb{N}$, $n^4 - n^2$ is a multiple of 12.

Exercise 1.10 A statement, $P(k)$ is defined for all $k \in \mathbb{Z}$. We know that $P(l)$ is true. We also know that for all $n \in \mathbb{Z}$

$$P(n) \implies P(n+1),$$
$$P(n) \implies P(n-1).$$

Prove that $P(n)$ holds for all $n \in \mathbb{Z}$.

Chapter 2
Double Counting

2.1 Introduction

Often mathematicians want to develop formulas for sums of terms. As we have seen, induction is one way to establish the truth of such formulas. However, induction relies on foreknowledge of the formula which has to be derived or guessed in some other way first. Induction also does not assist with recall. An alternate way to find and derive many counting formulas is *double counting*. With this technique, we divide up a collection of objects in two different ways. The results of the two different divisions must agree and this can yield a formula for the more complicated one. Such approaches have the advantage that the technique for finding the formula is often more memorable than the formula itself.

2.2 Summing Numbers

We wish to find the sum of the first N numbers. Call this x_N. So

$$x_N = \sum_{j=1}^{N} j.$$

We can regard the sum as a triangle with 1 stone in the first row, two stones in the second, three in the third and so on. See Fig. 2.1. Now if we make a copy of the triangle and rotate it through $180°$, we get N stones in the first row, $N-1$ in the second, $N-2$ in the third and so on. Joining the two triangles together, we have N rows of $N+1$ stones. The total number of stones is $N(N+1)$. So

$$X_N = \frac{1}{2}N(N+1).$$

© Springer International Publishing Switzerland 2015
M. Joshi, *Proof Patterns*, DOI 10.1007/978-3-319-16250-8_2

Fig. 2.1 Stones arranged in
rows with one more stone in
each row together with the
same stones rotated

We could also give this proof algebraically. Before proceeding to the algebra, we
look at a special case. If we have 5 stones, then

$$X_5 = 1 + 2 + 3 + 4 + 5,$$
$$X_5 = 5 + 4 + 3 + 2 + 1,$$
$$2X_5 = (1 + 5) + (2 + 4) + (3 + 3) + (4 + 2) + (5 + 1).$$

Now the algebra, if we reverse the order of the sum, we get

$$x_N = \sum_{j=1}^{N} (N + 1 - j).$$

Adding the two expressions for x_N together,

$$2x_N = \sum_{j=1}^{N} (N + 1) = N(N + 1),$$

and the result follows.

A similar argument can be used for the sum of the first N odd numbers

$$y_N = \sum_{j=1}^{N} (2j - 1).$$

Either by rotating the triangle or by reversing the order of the sum, we have

$$y_N = \sum_{j=1}^{N} (2N + 1 - 2j).$$

So

$$2y_N = \sum_{j=1}^{N} 2N.$$

We conclude that

$$y_N = N^2.$$

As well as being the sum of the first N odd numbers, N^2 is also the sum of the first N numbers plus the first $N-1$ numbers. One way to see this is take an $N \times N$ grid of stones and see how many stones lie on each upwards sloping diagonal. For the first N diagonals, the jth diagonal has j stones. These contribute the sum of the first N numbers. After N, the length of the diagonals goes down by one each time. The second set therefore contribute the sum of the first $N-1$ numbers and the result follows.

2.3 Vandermonde's Identity

Suppose we have $m+n$ jewels of varying sizes. There are m rubies and n sapphires. How many different ways can we select r jewels to be placed on a bracelet? Clearly, the answer is

$$\binom{m+n}{r}.$$

Note that since the jewels are all of different sizes, two different selections of r jewels are essentially different. However, if we first think in terms of using k rubies and $r-k$ sapphires, we see that the answer is also

$$\sum_{k=0}^{r} \binom{m}{k}\binom{n}{r-k}.$$

(We take the binomial coefficient to be zero when the inputs are out of their natural range. For example, if $r > m$, there are zero ways to choose r rubies.) In conclusion, we have *Vandermonde's identity*:

$$\binom{m+n}{r} = \sum_{k=0}^{r} \binom{m}{k}\binom{n}{r-k}.$$

2.4 Fermat's Little Theorem

Fermat's little theorem states

Theorem 2.1 *If a is a positive integer and p is prime then p divides $a^p - a$.*

An elementary proof can be made using double counting.

Proof Consider strings of letters of length p. The letters are from the first a letters in the alphabet. (If $a > 26$, we add extra letters to the alphabet.) How many such strings are there? Order matters, so we have a choices in each slot and there are p slots, so we get a^p different strings.

Now consider the operation on these strings of chopping off an element at the end and reinserting it at the front. Call this T^1. We define T^j to be the result of applying T^1 j times. Clearly, T^p is the identity map. We give some examples when $p = 3$ and $a = 2$.

$$T^1(AAA) = AAA,$$
$$T^1(ABA) = AAB,$$
$$T^2(ABA) = BAA.$$

Two strings, x and y, are said to be in the same *orbit* if there exists j such that

$$T^j x = y.$$

Note that then

$$T^{p-j} y = x.$$

Also note that if x and y are in the same orbit, and y and z are in the same orbit then x and z are too. So being in the same orbit is an equivalence relation. (See Appendix B for further discussion of equivalence relations.) This implies that every string is in exactly one orbit.

If a string is all one letter, e.g. "AAA", then it is the only string in its orbit. There are a such strings and so a orbits of size 1.

Now suppose a string x has more than one letter in it. Consider the strings

$$x, Tx, T^2x, \ldots, T^{p-1}x.$$

These will all be in the same orbit and everything in x's orbit is of this form. If we keep going we just get the same strings over again since $T^p x = x$. There are at most p elements in these orbits then. We show that when p is prime there are exactly p elements. If there were less than p then for some $k < p$, we would have

$$T^k x = x.$$

This means that cutting off the last k elements and sticking them at the front does not change the string. We also have

$$x = T^k x = T^{2k} x = T^{3k} x = T^{4k} x = \ldots$$

This implies that x is made of p/k copies of the first k elements. However, p is prime so its only divisor are 1 and p. If $k = 1$ then we have a string of elements the same which is the case we already discussed. If $k = p$ then we are just saying that $T^p x = x$ which is always true.

So we have two sorts of orbits, those with p elements and those with 1 element. We showed that there are a of the second sort. Let there be m of the first sort. Since there are a^p strings in total, we have

$$a^p = mp + a,$$

So

$$pm = a^p - a.$$

This says precisely that p divides $a^p - a$. $\qquad\qquad\square$

This proof is due to Golomb (1956).

2.5 Icosahedra

An icosahedron is a polyhedron with 20 faces. When working with three-dimensional solids, we can divide their surfaces into *vertices, edges* and *faces*. A vertex is a corner, a face is a flat two-dimensional side and an edge is the line defining the side of two faces.

A regular icosahedron is a Platonic solid. Every face is a triangle and the same number of faces meet at each vertex. How many edges does an icosahedron have? Call this number E. We know that there are 20 faces and that each face is a triangle. Define an edge-face pair to be a face together with one of the sides of the face which is, of course, an edge of the icosahedron. There are 60 such pairs since each face has 3 sides.

Each edge of the icosahedron lies in precisely two sides. The number of edge-face pairs is therefore double the number of edges. That is

$$2E = 60$$

and so $E = 30$.

2.6 Pythagoras's Theorem

The reader will already be familiar with the theorem that for a right-angled triangle, the square of the hypotenuse is equal to the sum of the squares of the other two sides. So if the sides are a, b and c, with c the longest side, we have

$$c^2 = a^2 + b^2.$$

We can use an extension of the double counting pattern to prove this theorem. Instead of using equal numbers of objects, we use equal areas. We take the triangle and fix a square with side length c to its side of that length. We then fix a copy of the triangle to each of the square's other sides. See Fig. 2.2. At each vertex of the square, we get 3 angles, each of which is one of the angles of the triangle so these add up to 180° and make a straight line. We now have two squares: the big one has side $a + b$ and the small one c. The former's area is

$$(a + b)^2 = a^2 + 2ab + b^2.$$

We can also regard the big square as the small one plus 4 copies of the triangle and so it has area

$$c^2 + 4 \times 0.5ab = c^2 + 2ab.$$

Fig. 2.2 Four right-angled triangles with sides a, b, c, placed on a square of side c

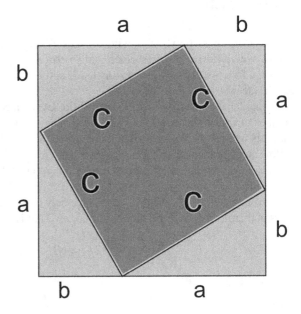

Equating these two, we get

$$a^2 + b^2 = c^2,$$

as required. Our proof is complete.

2.7 Problems

Exercise 2.1 Let a and b be integers. Develop a formula for

$$\sum_{j=1}^{n} a + jb.$$

Exercise 2.2 How many vertices does an icosahedron have?

Exercise 2.3 If p is a prime and not 2, and a is an integer, show that 2 divides into $a^p - a$. (Try to construct a bijection on the set of size p orbits which pairs them.)

Exercise 2.4 Suppose we have three sorts of jewels and apply the arguments for Vandermonde's identity, what formula do we find?

Chapter 3
The Pigeonhole Principle

3.1 Introduction

A pigeonhole is another name for a mailbox. If we have more pieces of mail than pigeonholes, someone's pigeonhole gets two pieces of mail. That is the pigeonhole principle. Whilst easy to state and rather obvious, it is very useful. It is closely related to the concept of *cardinality* which generalizes the concept of the number of elements of finite sets.

More formally, let A and B be sets such that the cardinality of A is greater than that of B. (i.e. A has more elements than B.) Suppose a function f maps from A to B then f is not injective that is there exists $x, y \in A$ such that

$$f(x) = f(y).$$

In this chapter, we look at a variety of applications from some quite different areas.

3.2 Rationals and Decimals

The reader will be familiar with the fact that some rational numbers are not easy to represent with decimals. In particular, their expansions repeat. For example,

$$\frac{1}{3} = 0.3333333\ldots$$

and

$$\frac{1}{9} = 0.111111\ldots$$

© Springer International Publishing Switzerland 2015
M. Joshi, *Proof Patterns*, DOI 10.1007/978-3-319-16250-8_3

More interestingly,

$$\frac{1}{7} = 0.142857142857142857142857\ldots$$

which repeats every 6 places.

Two obvious questions arise:

- If a decimal expansion repeats, will it always represent a rational number?
- If a number is rational, will its expansion always be finite or repeating?

The first of these is easy. Suppose the expansion of x repeats every k places after some point in the expansion. Then $y = 10^k x - x$ is zero after that point in the expansion. So we can make y into an integer simply by multiplying it by some power of 10. We then have for some $l, m \in \mathbb{N}$

$$10^l(10^k - 1)x = m.$$

Equivalently,

$$x = \frac{m}{10^l(10^k - 1)},$$

so x is rational.

For the second question, we analyze the algorithm that produces the terms in the decimal expansion. We have $q = m/n$, with m and n natural numbers. We take $q > 0$, since the result will follow for negative q simply by putting a minus sign in front.

If we proceed using the arithmetic algorithms we learnt in school, then eventually, we get to a point in the decimal expansion computation where we take a number r_0 and write

$$r_0 = a_0 n + b_0,$$

with $0 \leqslant b_0 \leqslant n - 1$. (See Chap. 4 for more discussion of the division algorithm.) The number a_0 is the term in the decimal expansion and b_0 is the remainder we use for the next term. We then put $r_1 = 10b_0$, and set

$$r_1 = a_1 n + b_1,$$

with $0 \leqslant b_1 \leqslant n - 1$. If we get a zero remainder the expansion terminates and we have a finitely long decimal. Otherwise, we keep going, setting $r_j = 10b_{j-1}$, and

$$r_j = a_j n + b_j.$$

If for some j, k with $j < k$, $b_j = b_k$ then the expansion starting at place k is the same as the one starting at place j. This is because the algorithm translated from j to k has the same inputs and the same operations.

However, if we consider the values

$$b_0, b_1, \ldots, b_{n-1},$$

there are n of these. If any is zero, the expansion terminates. If none is zero then they take values from 1 to $n - 1$. The pigeonhole principle then guarantees that two of them are equal. So for some j, k with $j < k \leqslant n - 1$,

$$b_j = b_k,$$

and the decimal repeats with period $k - j$. Note that the maximum period of repetition is, in fact, $n - 1$.

A consequence of our results is that we can characterize irrational numbers as those that have infinite non-repeating expansions. Note also that if a number has a finite or repeating expansion for one number base then it will have one of the two for all number bases.

3.3 Lossless Compression

I once saw an advertisement for a compression program that claimed to reduce the size of any file by at least 20%. This did not seem very believable so I asked a colleague if this was really possible. His response was that he doubted that it could compress the same file twice!

We can use the pigeonhole principle to prove that an algorithm that losslessly compresses any file is impossible. By lossless compression, we mean that it is possible to recover the original file purely from the data in the new file. For compression, the new file has to be smaller than the old file. A file in a modern digital computer is no more and no less than a sequence of 1 and 0 s. The amount of memory used is the length of the sequences.

A compression algorithm maps the sequences of length m to a sequence of length n with $m > n$. There are 2^m sequences of length m and 2^n sequences of length n. If we fill out any sequences of length less than n using zeros to be of length $m - 1$, then a lossless compression algorithm would give an injection from those of length m to those of length $m - 1$. However, a map from the sequences of length m to those of length $m - 1$ cannot be injective by the pigeonhole principle. In other words, it must be the case that two files get mapped to the same target file, and the decompression algorithm will not be able to say which is which. Lossless compression cannot occur.

However, most people use lossless compression programs all the time. So what is going on? In practice, most files, that have not already been compressed, have some structure that the compression program can work with. For example, a musical recording will display quite different characteristics from a text file or an image file. By recognizing this internal structure, the compression program can exploit it and compress the data. In particular, different compression formats are tuned to different sorts of files. We use MP3 files to store our music but not for our images.

3.4 More Irrationality

Let $x > 0$ be an irrational number. Consider the set

$$S_x = \{\mathrm{frac}(nx) \text{ for } n \in \mathbb{Z}\}.$$

Here $\mathrm{frac}(y)$ is the fractional part of y, that is the part obtained after discarding the whole number part: the bit after the decimal point if you prefer. More formally, it is the result after subtracting the largest integer less than or equal to y so $\mathrm{frac}(y)$ is always in the range $[0, 1)$. We use the pigeonhole principle to show that S_x contains a number arbitrarily close to zero: we show that if $\delta > 0$, then there exists $z \in S_x$ such that

$$0 < z < \delta.$$

All elements of S_x are positive, since if there was a zero the number x would be rational. So $S_x \subset (0, 1)$. Now suppose we pick n such that $n\delta > 1$. If we consider the intervals $(j/n, (j + 1)/n]$ for $j = 0, \ldots, n - 1$, there are n such intervals. This means that if we take the first $n + 1$ points in S_x, at least two must lie in one of these intervals by the pigeonhole principle. We therefore have values l and m with $m > l$, such that

$$|\mathrm{frac}(lx) - \mathrm{frac}(mx)| < 1/n < \delta.$$

In other words, there exist integers a, b and j such that

$$lx \in (a + j/n, a + (j + 1)/n), \quad mx \in (b + j/n, b + (j + 1)/n).$$

Now consider $(m - l)x = mx - lx$. This number is certainly positive and it is clearly in the interval

$$(b - a - 1/n, b - a + 1/n).$$

If it is bigger then $b - a$, its fractional part is less than $1/n$ and we are done. Otherwise, consider $(l - m)x$. This will lie in the range

$$(a - b, a - b + 1/n),$$

and taking its fractional part we are done. The crucial part of this argument was that the pigeonhole principle implied that eventually the sequence of numbers $\mathrm{frac}(nx)$ had to bunch together.

3.5 Problems

Exercise 3.1 Show that if a lossless compression algorithm exists for any file, then it is possible to represent any file by a single bit of information.

Exercise 3.2 Show that there are two residents of London with the same number of hairs on their heads.

Exercise 3.3 Suppose N people are at a party. Some have met before and some have not. Show that there are two people who have met the same number of other people before.

Exercise 3.4 Suppose we take a set of 101 different integers between 1 and 200. Show that there is a pair such that one divides the other.

Exercise 3.5 Represent the following decimals as ratios of integers.

- $0.1212121212\ldots$,
- $0.123123123123123\ldots$,
- $0.456456456\ldots$.

Chapter 4
Divisions

4.1 Introduction

In this chapter, we explore some basic results about division. These allow us to illustrate some proof techniques. The first result we want to prove is sometimes called the Division Lemma—it says that we can always write the answer as a whole number plus a remainder just as we did in primary school! To prove a result, we first have to formulate it correctly. Once that is done, we will see how it follows from more basic properties of natural numbers and, in particular, the *well-ordering principle.*

Our second task is to show that if two numbers, m and n, have highest common factor h then there exist integers a and b such that

$$am + bn = h.$$

We will show this via *algorithmic construction.* That is we construct an algorithm whose output is the numbers a, b and h and we prove that it always terminates.

4.2 Division and Well-Ordering

We have a natural number m and we want to divide it by another integer n. If we work solely with whole numbers, then if we proceed as we did when first learning arithmetic, we write

$$m = nq + r$$

with q, r natural numbers, and $0 \leqslant r < n$.

The Division Lemma states that such a decomposition is always possible. The standard way to prove it is to deduce it from the well-ordering of the natural numbers.

© Springer International Publishing Switzerland 2015
M. Joshi, *Proof Patterns*, DOI 10.1007/978-3-319-16250-8_4

We previously mentioned well ordering in the context of induction in an informal manner, now we give a formal definition.

Definition 4.1 A set E with an ordering $<$ is said to be well-ordered if every non-empty subset, F, of E has a smallest element. In other words, there exists $f \in F$, such that for all $g \in F$, we have $f \leqslant g$.

The standard ordering of the natural numbers is a well-ordering. We will take this for granted for now.

How does well-ordering help us? Let

$$F = \{m - nq \mid q \in \mathbb{Z}, \ m - nq \geqslant 0\}.$$

Note that if $x \in F$ and $x \geqslant n$ then $x - n \in F$ also. So any element bigger than $n - 1$ is not the smallest element of F. The set F is non-empty since it contains m, so by the well-ordering principle it has a smallest element. Let r be the smallest element of F. It must be smaller than n, since any element greater than or equal to n is not the smallest.

We therefore have

$$m = nq + r$$

with $0 \leqslant r < n$, as required.

4.3 Algorithms and Highest Common Factors

The highest common factor of two natural numbers, m and n, is the largest natural number, h, which divides into both of them with no remainders. This definition presupposes that such a number exists! It is, however, easy to show that there is such a number. First, the set of factors is non-empty since it contains 1. Second, any factor of a number is less than or equal to it. So the largest element is less than or equal to $\min(m, n)$. We therefore have a finite set of common factors and its largest element is the highest common factors. (Note the implicit use of well-ordering here—we are using that a finite non-empty subset of the integers has a largest element. How would you prove this?)

Our proof of the existence of the highest common factor is not very constructive—it does not tell us how find it. Indeed, the implicit method is to test every number from 1 to $\min(m, n)$ and see if it divides both m and n. Once this is done, take the largest such number. Whilst that algorithm would work, it is not very efficient. How could we improve it? One simple improvement is simply to count downwards and stop as soon as a common divisor is reached. This is still not very efficient, however.

Fortunately, there is a much better method known as the *Euclidean algorithm*. We will write (m, n) for the highest common factor from now on to simplify notation. It relies on the observation that if

$$m = qn + r$$

with m, n, q, r all counting numbers then

$$(m, n) = (n, r).$$

To see this, note that we can also write

$$r = m - qn.$$

If x divides a and b then it also divides $a + b$ and $a - b$. Setting $a = qn$ and $b = r$, any factor of r and n is a factor of m. With $a = m$ and $b = qn$, any factor of n and m is also a factor of r. Since the set of common factors of m and n is the same as the set of common factors of n and r, the two pairs must have the same highest common factor.

There is an obvious choice for r and q: the results of the Division lemma. This will make r as small as possible. If r is zero, then we have

$$m = qn$$

which means that the highest common factor is n. Otherwise, we can repeat, and

$$n = q_1 r + r_1,$$

with $r_1 < r$. We have
$$(m, n) = (n, r) = (r, r_1).$$

Letting $r_0 = r$, we can now keep going and obtain a sequence of remainders r_j. We terminate when we obtain a zero remainder. As argued above, we have

$$(m, n) = (r_j, r_{j+1})$$

for all j with $r_{j+1} > 0$.

Once we hit a point with zero remainder, the algorithm terminates. We need to show that it does. However, we always have

$$r_{j+1} < r_j,$$

so it must terminate in at most n steps.

We do an example with $m = 57$ and $n = 51$.

$$57 = 1 \times 51 + 6,$$
$$51 = 8 \times 6 + 3,$$
$$6 = 2 \times 3 + 0.$$

The highest common factor is 3.

A consequence of the Euclidean algorithm is that we can always express a highest common factor as an integer combination of the two original numbers:

Theorem 4.1 *If m and n are counting numbers then there exist integers a and b such that*

$$(m, n) = am + bn.$$

In our last example,

$$3 = 51 - 8 \times 6 = 51 - 8 \times (57 - 51) = 9 \times 51 - 8 \times 57.$$

We simply work our way backwards through the algorithm.

We can formally prove this theorem using complete induction. We assume $n \leqslant m$ without loss of generality since they can always be switched. First note that if n is the highest common factor then the result holds

$$n = 1 \times n + 0 \times m.$$

Otherwise, let r_j be the remainder after $j + 1$ steps. For convenience, let $r_{-1} = n$. The first remainder r_0 is certainly an integer combination since

$$r_0 = m - qn.$$

Now assume the result holds for r_l for $l \leqslant j$. We have

$$r_{j-1} = q_j r_j + r_{j+1},$$

and so

$$r_{j+1} = r_{j-1} - q_j r_j$$

Substituting the linear combinations for r_{j-1} and r_j, we have a linear combination for r_{j+1} and the result follows.

Our proof proceeded by using an algorithm that constructed the numbers a and b. We also proved that the algorithm does terminate. This is an example of *algorithmic construction*.

4.4 Lowest Terms

The reader will have reduced fractions to lowest terms many times at school. What does that mean mathematically and how can we show that it is always possible?

Theorem 4.2 *Suppose p and q are counting numbers then there exists counting numbers p_1 and q_1 such that*

$$(p_1, q_1) = 1, \text{ and } \frac{p}{q} = \frac{p_1}{q_1}.$$

Proof Let $m = (p, q)$. We can then set $p_1 = p/m$ and $q_1 = q/m$. Clearly,

$$\frac{p}{q} = \frac{mp_1}{mq_1} = \frac{p_1}{q_1}.$$

Let $k = (p_1, q_1)$ we have that km divides p and q. Since m is their highest common factor, k must be 1 and we are done. □

4.5 Euclid's Lemma

We can now use the result of the last section to prove a result about factors which is sometimes called Euclid's Lemma.

Lemma 4.1 *Suppose k, m, and n are counting numbers such that $(k, m) = 1$ and k is a factor of mn, then k is a factor of n.*

Proof We have that there exists integers a and b such that

$$am + bk = 1.$$

So

$$amn + bkn = n.$$

We have that mn is a multiple of k so amn is. So

$$amn = \alpha k,$$

for some α. Clearly, bkn is nb times k. So

$$n = (\alpha + bn)k$$

and k is a factor of n. □

A nice consequence of this lemma is

Corollary 4.1 *Let m and n be positive natural numbers. If p is a prime or 1, and $p|mn$ then p divides at least one of m and n.*

Proof If p is 1 it divides anything, so assume that p is prime. If p divides m then we are done. Otherwise, the highest common factor of m and p is 1, since the highest common factor must divide into p and its only factors are p and 1. Euclid's lemma then states that p divides n and we are done.   □

Note that many authors state this lemma with the hypothesis that p is prime rather than that p is prime or 1. The more general version is certainly true and it will be convenient later when we are trying to prove that a certain number is 1 to allow its possibility here.

4.6 The Uniqueness of Prime Decompositions

We know from Sect. 1.2 that every natural number bigger than one can be written as a product of positive integer powers of prime numbers. So given m, there exists p_j prime and $\alpha_j \in \mathbb{N}$, $\alpha_j > 0$, such that

$$m = p_1^{\alpha_1} p_2^{\alpha_2} \cdots p_k^{\alpha_k}.$$

Rearranging if necessary, we can assume that $p_i < p_{i+1}$ for all i. We want to show that this representation is unique. Suppose we also have primes q_j and powers β_j with the same properties. We need to show that

$$q_j = p_j$$

for all j and $\alpha_j = \beta_j$.

If p and q are prime numbers they are either equal or coprime. We have that for each l

$$q_l | p_1^{\alpha_1} p_2^{\alpha_2} \cdots p_k^{\alpha_k}.$$

From Corollary 4.1, we have that q_l divides at least one of

$$p_1^{\alpha_1} \quad \text{and} \quad p_2^{\alpha_2} \cdots p_k^{\alpha_k}.$$

Repeating the argument, we see that for some r,

$$q_l | p_r^{\alpha_r}.$$

If $\alpha_l > 1$, we can argue in the same way that q_l divides p_r or $p_r^{\alpha_r - 1}$. Repeating, we have

$$q_l | p_r$$

so $p_r = q_l$.

Since the problem is symmetric in the ps and qs, for every p_r there exists an l such that $p_r = q_l$. In other words, we have the same sets of primes.

It remains to show that the powers are the same. We need to show that if

$$p_1^{\alpha_1} p_2^{\alpha_2} \cdots p_k^{\alpha_k} = p_1^{\beta_1} p_2^{\beta_2} \cdots p_k^{\beta_k},$$

with p_j prime and α_l, β_l positive natural numbers then $\alpha_l = \beta_l$ for all l.

For each point where $\alpha_l \geqslant \beta_l$, we divide both sides by $p_l^{\beta_l}$. For any remaining values of l, we divide both sides by $p_l^{\alpha_l}$. We then have primes to the power $\alpha_l - \beta_l$ on the left hand side and primes to the power $\beta_l - \alpha_l$ on the right hand side. The sets of primes on each side with non-zero powers are disjoint.

If we repeat the argument above where we showed that the primes on each side must be the same, we realize that all the powers must be zero. In others $\alpha_l = \beta_l$ for all l and we are done.

4.7 Problems

Exercise 4.1 Find the highest common factors of the following number pairs.

- 1236 and 369.
- 144 and 900.
- 99 and 36.

Exercise 4.2 Show that the highest common factor of n and $n + 1$ is 1.

Exercise 4.3 Show that the highest common factor of n and $n^2 + 1$ is 1.

Exercise 4.4 What are the possible highest common factors of n and $n^2 + k$ if $k < n$?

Chapter 5
Contrapositive and Contradiction

5.1 Introduction

Two very common proof tools are proof by contrapositive and proof by contradiction. Whilst these are related, they are distinct notions and it is useful to distinguish between them. They are both really principles of logic rather than of mathematics.

In what follows let P, Q and R be mathematical statements that may be true or false. We write $\neg P$ for "not P" and so on. We write $P \& Q$ for the statement that both P and Q hold. Proof by contrapositive states that the following two statements are logically equivalent:

$$P \implies Q;$$
$$\neg Q \implies \neg P.$$

If we can prove one of them, then the other holds. If we can disprove one, then the other is false.

Proof by contradiction involves a third statement, R. It states that if there is a statement R such that

$$P \& \neg Q \implies R, \text{ and}$$
$$P \& \neg Q \implies \neg R$$

then

$$P \implies Q.$$

© Springer International Publishing Switzerland 2015
M. Joshi, *Proof Patterns*, DOI 10.1007/978-3-319-16250-8_5

5.2 An Irrational Example

Many proofs by contrapositive are phrased as proofs by contradiction. The objective is to prove

$$P \implies Q$$

and they are structured as follows:

(1) assume P is true;
(2) assume $\neg Q$ is true;
(3) show that $\neg Q$ implies $\neg P$;
(4) deduce that $\neg P$ is true;
(5) since we have that both P and $\neg P$ are true, we have shown assuming $\neg Q$ is true leads to a contradiction so Q is true;
(6) that is we have shown $P \implies Q$.

First, note that in this proof by contradiction the third statement, R, is actually P. Second, the crucial part of this proof structure is line 3. This may or may not use the truth of P which was assumed at step 1. If it does not then we can short-circuit the argument and replace it with

(1) show that $\neg Q$ implies $\neg P$;
(2) invoke the equivalence of contrapositives and deduce that $P \implies Q$.

The second structure is a proof by contrapositive rather than contradiction and it is much simpler when it applies.

A standard well-known example of proof by contradiction is the irrationality of the square root of 2. The argument goes as follows.

- Suppose $\sqrt{2}$ is rational then there exists m, n positive natural numbers such that $2 = m^2/n^2$ where m and n have no common factors.
- We can write $m^2 = 2n^2$.
- So m^2 is even. The number m is even or odd. The square of an odd number is odd so m is not odd.
- So m is even and $m = 2k$ for some integer k.
- $2n^2 = 4k^2$ so $n^2 = 2k^2$.
- So n^2 is even and by the same argument as above, n is even.
- So 2 is a common factor of m and n, but they were said to have no common factors so we have a contradiction.
- Our initial assumption must be false and $\sqrt{2}$ is not rational.

Some points to note, if we phrase our statements correctly, we do not need $\sqrt{2}$ to exist as a real number for this proof; the statement that lead to a contradiction was

$$2 = (m/n)^2$$

not that $\sqrt{2} = m/n$. The difference being that the latter statement requires us to work in a bigger number system where $\sqrt{2}$ makes sense. Second, we used the fact that any rational number can be expressed as a ratio of two integers that have no common factors which we proved as Theorem 4.2.

Before considering contrapositives, it is worth thinking about what we have proven and to what extent the result is generalizable. Our main result is that

$$x^2 = 2,$$

has no rational solutions. A natural question to ask is "what is special about 2?" Of course, many things are special about 2. It is even, it is prime, and it appears twice in this equation. The crucial argument in the proof was that if 2 divided the square of a number, m, then 2 also divides m. So we can expect the same result to hold for the square root of any number k that has this property. Prime numbers have this property, see Sect. 4.5:

$$p | m^2 \implies p | m, \qquad (5.2.1)$$

so it will follow that the square root of any prime is also irrational once we have proven (5.2.1). We do this below.

Perfect squares trivially have rational square roots. However, there are many numbers that are neither prime nor are perfect squares. The smallest such number is 6, another one is 12. If we use a calculator to compute their square roots, these show no obvious patterns and so might be irrational. My computer gives

$$\sqrt{6} = 2.44948974278318 \text{ and } \sqrt{12} = 3.46410161513775.$$

Remember that computer numbers are always approximations; this does not mean that there are not more digits. We certainly have

$$12 \,|\, 36$$

but 12 does not divide into 6. So our proof does not generalize to 12. Of course, that does not mean that $\sqrt{12}$ is not irrational, merely that we need a different method of proof.

What if we proceed via contrapositive? Our new statement is: if q is rational and not an integer then q^2 is not an integer. If we can we prove this, then we have shown that all non-perfect squares have irrational square roots. Even more generally, we could consider the analogous statement for kth powers: for $k \geqslant 2$, if q is rational and not an integer then q^k is not an integer. Once this is proven, we will have that all kth roots are integers or irrational.

Now for the proof: if q is rational and not an integer then q can be written as m/n with m and n coprime and $n > 1$ using Theorem 4.2. Note that that if $n = 1$, q is an integer which we assumed was not the case so $n \neq 1$. Now consider

$$q^k = \frac{m^k}{n^k}.$$

This will be an integer if and only if n^k divides into m^k. We show that it cannot. We show that no natural number bigger than 1 divides both n^k and m^k. This will follow if we show no prime number divides both numbers, since any natural number greater than 1 is prime or composite, and any composite number will have a prime factor (see below). Let p be a number that is either 1 or a prime such that

$$p|m^k \text{ and } p|n^k.$$

We need a lemma:

Lemma 5.1 *Let l be a positive integer. If a number, p, that is 1 or prime divides l^k for $k \geqslant 2$ then $p|l$.*

Given this lemma, we have that p divides m and n, but m and n have highest common factor 1 so p must be 1. So q^k is not an integer and every root of an integer that is not an integer is irrational.

We still have to prove our lemma.

Proof We use Corollary 4.1. If p is 1 there is nothing to prove, so assume that p is prime. We use induction. If $k = 2$, then this is just Corollary 4.1. For general k assume the result is known for k and suppose $p|l^{k+1}$ then writing

$$l^{k+1} = l^k l,$$

we have by the same Corollary that either $p|l$ or $p|l^k$. In the first case, we are done, and in the second, the result follows immediately from the inductive hypothesis and we are done.   □

Although the irrationality of the square root is a standard example of the power of proof by contradiction, we have seen that is perfectly possible to prove it by contrapositive. We have also proven a much more general result: any kth root of a positive integer that is not an integer is irrational.

5.3 The Infinitude of Primes

There are an infinite number of primes. Before proceeding to the proof. We first establish a lemma.

Lemma 5.2 *Every integer, y, bigger than 1 has a prime factor.*

Proof If y is prime, we are done. Otherwise, $y = k_1 l_1$ with $k_1, l_1 > 1$. Note that $k_1 \leqslant y/2$. Either k_1 is prime and we are done, or $k_1 = k_2 l_2$, with $k_2, l_2 > 1$. We now have

$$k_2 \leqslant k_1/2 \leqslant k_1/4.$$

Either k_2 is prime or we can construct $k_2 = k_3 l_3$ and so on. Since $2^y \geqslant y$, this process must halt by step y and we are done.

One could also prove this result using complete induction—in fact, it follows directly from the results of Sect. 1.2.   □

How can we prove the infinitude of primes?

First, we use proof by contradiction.

Proof Suppose there are a finite number, N, of primes. Then we can label them, $p_1, p_2, \ldots, p_N$ to make a list. If we now let

$$x = p_1 p_2 p_3 \ldots p_N + 1,$$

then x is either prime or composite. It cannot be prime since it is not on the list. We also have that x divided by p_j has remainder 1 for each j so it is not a product of the primes on our list either. However, by our lemma x has a prime factor p, but p is not on our list. We have a contradiction: assuming the list was finite and complete led to the existence of a prime not on the list. So our initial assumption that a finite list could be complete was false, and there are an infinite number of times.   □

However, we do not really need proof by contradiction.

Proof We show that any finite list of primes is not complete, and so the number of primes must be infinite. Let $p_1, p_2, \ldots, p_N$ be a list of primes. Let

$$x = p_1 p_2 p_3 \ldots p_N + 1.$$

From our lemma, x has a prime factor, p. Each p_j does not divide x, since the remainder is 1. So $p_j \neq p$ for all j. We have constructed a prime not on the list, and so the list is not complete, as claimed. The number of primes is therefore infinite. □

The two arguments are very similar. What is the difference? In the first, we assume that the list of ALL primes is finite and show that that generates a contradiction. In the second, we show that any given finite list of primes is not complete. This proves that the number of primes is greater than N for all N and so that it is infinite.

5.4 More Irrationalities

Suppose x is irrational and q is rational. What can we say about $y = x + q$? If y is rational, then we have for some integers, k, l, m, n,

$$\frac{k}{l} = x + \frac{m}{n}.$$

It follows that

$$x = \frac{k}{l} - \frac{m}{n} = \frac{kn - lm}{ln}.$$

The right hand side is a rational number so this contradicts the irrationality of x. Invoking proof by contradiction, we conclude that y is not rational.

Alternatively, we could employ proof by contrapositive. Our contrapositive is

Proposition 5.1 *If q_1 and q_2 are rational and for some $x \in \mathbb{R}$, we have*

$$q_1 = q_2 + x,$$

then x is rational.

Proof We have

$$x = q_1 - q_2 = \frac{k}{l} - \frac{m}{n} = \frac{kn - lm}{ln}$$

for integers k, l, m and n. So x is rational. $\qquad\qquad\qquad\qquad\qquad\square$

If x is irrational and q_2 is rational, we therefore must have that by contrapositive that q_1 is irrational and we are done.

5.5 The Irrationality of e

Recall that the number e is defined by

$$e = \sum_{n=0}^{\infty} \frac{1}{n!}.$$

When studying e it will be convenient to use some results about geometric series. The main result we will use is

Proposition 5.2 *If $|x| < 1$, then $\sum_{j=0}^{\infty} x^j$ exists and equals $(1 - x)^{-1}$.*

With this fact handy, we first look at why the sum for e makes sense. Let

$$x_k = \sum_{n=0}^{k} \frac{1}{n!}.$$

The sequence x_k is increasing since each additional term is positive. Any bounded increasing sequence converges (see Chap. 10). So we need to show that the terms x_k are bounded. We use that for $n > 1$,

$$\frac{1}{n!} \leqslant \left(\frac{1}{2}\right)^{n-1}.$$

To see this just replace each number bigger than 2 in the definition of $n!$ with 2. So

$$x_k \leqslant 2 + \sum_{n=2}^{k} \left(\frac{1}{2}\right)^{n-1}.$$

So $x_k < 4$ for all k by our result on geometric series. We have a bounded increasing sequence and convergence follows. We have given an example of proof of convergence by *domination*. To prove that a series of positive terms converges, we find another convergent positive series of which each individual term is bigger.

We have shown that the number e makes sense. We still need to show that it is irrational. We show that for any natural number q, the number qe is not an integer. Any rational number can be turned into a integer simply by multiplying by its denominator so this is enough. If qe is an integer, then so is $q!e$. So it is enough to show that $q!e$ is not an integer for any $q \in \mathbb{N}$.

We can write

$$q!e = \sum_{k=0}^{q} \frac{q!}{k!} + \sum_{k=q+1}^{\infty} \frac{q!}{k!}.$$

If $k \leqslant q$,

$$\frac{q!}{k!} = (k+1)(k+2)\ldots(q-1)q,$$

and this is an integer. We will show that the remaining term is between zero and 1 which proves that $q!e$ is not an integer. For $k > q$,

$$\frac{q!}{k!} = \frac{1}{q+1}\frac{1}{q+2}\cdots\frac{1}{k} \leqslant \left(\frac{1}{q+1}\right)^{k-q}.$$

Using the sum of a geometric series with $x = 1/(q+1)$, we now observe

$$\sum_{m=1}^{\infty} \left(\frac{1}{q+1}\right)^{m} = \frac{1}{q}$$

which is less than 1. This shows that the tail of the expansion of $q!e$ is less than 1. Since it is between zero and one, it is not a whole number. It follows that $q!e$ is not a whole number and so that e is not rational.

5.6 Which to Prefer

Some mathematicians love proof by contradiction. Some hate it. I generally find that I can rephrase proofs by contradiction to be proofs by contrapositive, and I generally find such rephrasings cleaner. However, it is often easier to find the proof by contradiction in the first place. Ultimately, mathematicians greatly prefer any proof to no proof! However, it is generally worthwhile to see whether a proof can be rephrased if for no other reason than that the process of rephrasing yields extra insights.

The main reason I prefer proofs by contrapositive is that everything within the proof is true under the contrapositive assumptions. With proof by contradiction, your objective is to show that everything after the assumptions is false! This means that with proof by contrapositive, I get to know many additional things I do not get with proof by contradiction.

5.7 Contrapositives and Converses

I wish to emphasize that the contrapositive is not the converse. The contrapositive of

$$P \implies Q$$

is

$$\neg Q \implies \neg P.$$

The converse is

$$Q \implies P$$

which has contrapositive

$$\neg P \implies \neg Q.$$

Whilst statements are logically equivalent to their contrapositives, converses are a different matter. A statement can be false and its converse true and vice versa. Suppose n is a number bigger than 2. The statement n is prime implies that n is odd is true. However, if n is odd it does not have to be prime.

5.8 The Law of the Excluded Middle

Another basic technique from logic is the *law of the excluded middle*. It states that given a statement P, either P is true or not P is true. No other possibility can occur. We give a brief example of its application. We want to prove that there exist irrational numbers p and q such that

$$p^q$$

is rational. Consider

$$\left(\sqrt{2}^{\sqrt{2}}\right)^{\sqrt{2}} = \sqrt{2}^2 = 2.$$

This number is certainly rational. Let P be the statement that

$$\sqrt{2}^{\sqrt{2}}$$

is irrational. Either P is true and so setting

$$p = \sqrt{2}^{\sqrt{2}}, \quad q = \sqrt{2},$$

we have our example. Or P is false and we have that

$$\sqrt{2}^{\sqrt{2}}$$

is rational, and we are done. We have proven the existence of a pair (p, q) both irrational such that p^q is irrational. However, we have not found such a pair! Our proof is silent on which of the two possibilities holds. The law of the excluded middle, like proof by contradiction, is not constructive.

5.9 Problems

Exercise 5.1 Use proof by contradiction to show that $\sqrt{12}$ is irrational.

Exercise 5.2 What are the contrapositives of the following statements?

- Every differentiable function is continuous.
- Every infinite set can be placed in a bijection with the rationals.
- Every prime number is odd.
- Every odd number is prime.

What are their converses? Which are true?

Chapter 6
Intersection-Enclosure and Generation

6.1 Introduction

In many areas of mathematics, we study subsets with specific properties. For an arbitrary subset, we then want to find its smallest superset with the requisite properties. The first question to answer is "does such a smallest set exist?" Thus we have a set A contained in some larger set E. We have a property P and we want to find a set B such that

$$A \subseteq B \subseteq E$$

which has the property P, and which is contained in all subsets of E which both contain A and have the property P. One solution to this problem is to use the *intersection-enclosure* pattern.

6.2 Examples of Problems

A non-empty subset, B, of $\mathbb{R}$ is said to be an *additive subgroup* if the difference of any two elements of B is also in B. In symbols

$$x, y \in B \implies x - y \in B.$$

Note that $0 = x - x \in B$. It follows that $-x$ is also in B. A quick consequence is that $x + y$ is in B if x and y are in B, as well as $x - y$.

Our problem is to show that for any set A there is a smallest additive subgroup containing it.

A trivial example is "what is the smallest additive sub-group containing 1?" The answer is the integers! Try to convince yourself that is indeed the case. A slightly harder example is to let A be $\{\sqrt{2}\}$. We then end up with the integers scaled by $\sqrt{2}$.

© Springer International Publishing Switzerland 2015
M. Joshi, *Proof Patterns*, DOI 10.1007/978-3-319-16250-8_6

A non-empty subset, B, of $\mathbb{R}^n$ is said to be a *vector sub-space* if the sum of any two elements of B is also in B, and any scalar multiple of an element of B is also in B. In symbols

$$x, y \in B \quad \Longrightarrow \quad x + y \in B,$$
$$x \in B, \ \lambda \in \mathbb{R} \quad \Longrightarrow \quad \lambda x \in B.$$

The problem is therefore to show that there is a smallest subspace containing any set A.

We give some examples of sub-spaces. Consider any vector x. The set

$$B_x = \{\lambda x \mid \lambda \in \mathbb{R}\}$$

is a vector sub-space and is, in fact, the smallest one containing x. Another simple example is to consider all the vectors whose first coordinate vanishes. Adding two such vectors will result in a vector with zero first coordinate, and multiplying by a scalar will not change the first coordinate's zeroness.

Obviously, there is nothing special about the first coordinate, and we could obtain a subspace by choosing any one of the coordinates and requiring it to be zero. More generally, we could make a fixed subset of coordinates zero.

A subset B of $\mathbb{R}$ is said to be *closed* if the limit of every convergent sequence in B is also in B. Thus

$$x_n \in B \ \forall n, \ x_n \to x \in \mathbb{R} \quad \Longrightarrow \quad x \in B.$$

For example, the set $[0, \infty)$ is closed since the limit of a sequence of non-negative points is also non-negative. However, the set $B = (0, \infty)$ is not. If we take $x_n = 1/n$ then x_n converges to zero which is not in B.

The integers are also a closed subset of $\mathbb{R}$ since a sequence of integers only converges if it is eventually constant, and the limit is then this final constant.

A subset, B, of $\mathbb{R}^n$ is said to be *convex* if the straight line between any two of its points is also contained in it. So

$$x, y \in B, \ \lambda \in (0, 1) \quad \Longrightarrow \quad \lambda x + (1 - \lambda)y \in B.$$

6.3 Advanced Example

In probability theory, the concept of a *sigma algebra* is important. We have a set Ω and a collection of subsets of Ω typically called $\mathcal{F}$. This collection is called a sigma algebra if

- $\Omega \in \mathcal{F}$,
- $C \in \mathcal{F}$ implies $C^c \in \mathcal{F}$,

- $C, D \in \mathcal{F}$ implies $C \cup D \in \mathcal{F}$,
- if C_n is a sequence of sets in $\mathcal{F}$ then $\bigcup\limits_{n=1}^{\infty} C_n \in \mathcal{F}$.

The same problem arises of finding the smallest sigma algebra which contains a given collection of subsets.

6.4 The Pattern

In all the examples above, there is an important additional feature. If we take an intersection of subsets with a given property then the intersection also has the property. This intersection can be over an infinite collection of subsets and the property is still retained.

For example, if E_α is an additive sub-group of $\mathbb{R}$ for all $\alpha \in I$ for some index set I, then

$$E = \bigcap_{\alpha \in I} E_\alpha = \{x \mid x \in E_\alpha, \forall \alpha\},$$

is also an additive sub-group. To see this, if $x, y \in E$ then $x, y \in E_\alpha$ for all $\alpha \in I$ so $x - y \in E_\alpha$. Since this is true for all α, we have

$$x - y \in E$$

and E is indeed an additive subgroup.

The pattern is therefore to consider the set of subsets with the given property which contain a given set A and take their intersection. We need that the encompassing set E has the property to ensure that the intersection is over a non-empty collection. We thus let

$$B = \{x \mid x \in C \ \forall C \text{ such that } A \subseteq C \text{ and the property holds for } C\}.$$

First, B contains A since A is a subset of all the sets C which we intersect. Second, since the property is preserved by intersections and it holds for each set C, it also holds for their intersection, that is B.

We have thus constructed a set B containing A for which the property holds. We now need to show that it is the smallest such set. However, if we take an arbitrary set, C, with the property then it would have been part of the intersection. So B is its intersection with some other sets and by the definition of intersection must be contained in it.

In conclusion, B is the smallest set containing A for which the property holds. This construction works provided the property holds for the encompassing set E and is preserved by arbitrary intersections.

An intersection of vector subspaces is also a vector subspace, and the smallest one containing A is said to be *generated* by A or sometimes its *linear span*.

Closedness is preserved by intersections and the smallest closed set containing $A \subseteq \mathbb{R}$ is called its *closure* and generally written $\bar{A}$.

The smallest convex set containing $A \subseteq \mathbb{R}^n$ exists by the same argument, and is called the *convex hull*.

6.5 Generation

The set B produced from A by the enclosure-intersection pattern we have discussed is often said to have been *generated* by A. There is often an alternative way to find B which is more intuitive. However, proving that it works is sometimes harder. In this section, we explore this alternative.

Instead of considering intersections, suppose instead that we keep on adding in the elements that stop the property from holding. First, we look at additive subgroups of $\mathbb{R}$. We analyze the subset $\{1\}$ using generation. Let

$$A_0 = \{1\}.$$

It is not an additive subgroup because it does not contain $1 - 1 = 0$. So we add in zero. We therefore let
$$A_1 = \{0, 1\}.$$

We are now missing -1. So we let

$$A_2 = \{-1, 0, 1\}.$$

But we now need -2 and 2 and so on.

We can therefore let

$$A_{j+1} = A_j \cup \{x - y \mid x, y \in A_j\},$$

for $j = 1, 2, 3, \ldots$. The problem is that the process may not terminate. If it does then we have for some k
$$A_k = A_{k+1}$$

and this means that A_k is an additive subgroup. It is the smallest one containing A since any additive subgroup is required to be invariant under the operations used to generate A_k.

However, it is clear that for our example the process does not terminate. To see this, first observe that A_j has a finite number of elements for all j. It must therefore have a largest element, z. Clearly, $z + 1$ is in $A_{j+1} - A_j$.

A solution is at hand, however. We simply take the union, let

$$A_\infty = \bigcup_k A_k.$$

The set A_∞ contains A trivially. It is contained in the smallest additive subgroup, B, containing A since B must contain A_k for all k.

It remains to check that A_∞ is indeed an additive subgroup. If $x, y \in A_\infty$ then $x \in A_l$ for some l and $y \in A_m$ for some m. We then have

$$x, y \in A_{\max(l,m)}.$$

So

$$x - y \in A_{\max(l,m)+1}.$$

This shows that $x - y \in A_\infty$ and we are done.

This argument worked because the definition of an additive sub-group is essentially finite. We only look at the difference of two elements and require that to be in the same set. This allowed us to specialize down to a point before infinity. If our property was defined in terms of a larger finite number of elements then the argument would still work. However, if an infinite number of elements were involved then it might not.

What could we do if A_∞ did not have the invariance property required? One solution is simply to start over again using A_∞ in place of A_0. Thus we get a sequence of sets $A_{\infty,j}$ and we can define

$$A_{\infty,\infty} = \bigcup_k A_{\infty,k}.$$

If the property still does not hold, then there is nothing to stop us repeating as many times as we want. Indeed, we can do so infinitely often and then take the union of all the generated sets again. Hopefully, that will be enough but if not, we can keep going.

More generally, we can imagine an operation on our set that involves an uncountable number of its elements and this will be even harder. (We discuss the notion of uncountability in Chap. 13.) To give a general proof that this process of using bigger and bigger unions will eventually terminate is actually very hard. So whilst generation is in many ways more intuitive than intersection-enclosure, it can be harder to work with.

The reader may be curious to see an example where generation terminates before infinity, that is an example where $A_j = A_{j+1} \neq A_0$ for some j. Let

$$A_0 = \{2j \mid j \in \mathbb{Z}\} \cup \{1\}.$$

So A_0 is all even integers together with 1. We clearly do not have an additive subgroup and A_1 is the set of all integers, $\mathbb{Z}$. However, $\mathbb{Z}$ certainly is an additive subgroup and the process terminates.

6.6 Fields and Square Roots

The real numbers have various properties including closure under addition and multiplication, the existence of additive inverses and for non-zero elements the existence of multiplicative inverses. In equation terms,

$$x, y \in \mathbb{R} \implies x + y, xy \in \mathbb{R},$$
$$x \in \mathbb{R} \implies \exists y, \; x + y = 0,$$
$$x \in \mathbb{R} - \{0\} \implies \exists y, \; xy = 1.$$

An extra property that $\mathbb{R}$ has, but $\mathbb{Q}$ does not, is that non-negative numbers have square roots. So

$$x \in \mathbb{R} \implies \exists y, \; y^2 = |x|.$$

We can therefore write $\sqrt{|x|}$ for elements of $\mathbb{R}$. For elements of $\mathbb{Q}$ we can write $\sqrt{|x|}$ but we are not guaranteed that the result is in $\mathbb{Q}$.

The question then arises of finding the smallest subset of $\mathbb{R}$ closed under addition, subtraction, multiplication and division which contains $\mathbb{Q}$ and is also closed under the taking of square-roots of positive numbers. This is called finding the smallest sub-field with the square-root property.

First, does such a smallest sub-field exist? Yes. This is a direct application of intersection-enclosure. The set $\mathbb{R}$ contains $\mathbb{Q}$ and is invariant under these operations. Since these properties are all defined in a finite way, it is trivial to check that any intersections of such sets is also invariant so intersection-enclosure does apply.

The existence of a smallest sub-field does not imply that the smallest sub-field is not $\mathbb{R}$. In fact, there are smaller sub-fields but we have to prove they exist. Now suppose we apply generation. Let $\mathbb{F}_0 = \mathbb{Q}$. We let $\mathbb{F}_j$ be the set of points of the form

$$p + q\sqrt{r} \text{ for } p, q, r \in \mathbb{F}_{j-1}$$

with $r > 0$. Note that it would be equivalent to use $p + q\sqrt{|r|}$, since $r \in \mathbb{F}_j$ implies that $-r \in \mathbb{F}_j$. We then let

$$\mathbb{F} = \bigcup_j \mathbb{F}_j.$$

The finiteness of the conditions guarantees that $\mathbb{F}$ is indeed a square-root closed sub-field.

In fact, each set $\mathbb{F}_j$ is a sub-field. For example, if $x \in \mathbb{F}_j$ then

$$x = p + q\sqrt{r}, \text{ with } p, q, r \in \mathbb{F}_{j-1},$$

and

$$x^{-1} = \frac{1}{p + q\sqrt{r}} = \frac{p - q\sqrt{r}}{p^2 + q^2 r}$$

which is clearly in $\mathbb{F}_j$. The other operations are easily checked. However, for square-roots there is no reason to think that $\mathbb{F}_j$ is closed. But if $x \in \mathbb{F}$, then for some j, $x \in \mathbb{F}_j$ so $\sqrt{|x|} \in \mathbb{F}_{j+1} \subset \mathbb{F}$, so $\mathbb{F}$ is invariant under square roots. Similarly, since every $\mathbb{F}_j$ is closed under the other operations so is $\mathbb{F}$.

The question remains of whether $\mathbb{F} = \mathbb{R}$. We prove that it is not by showing that a well-known number, the cube-root of 2, is not in $\mathbb{F}$. We show

Theorem 6.1 *If the equation*

$$x^3 - m = 0$$

with m a rational has a solution in $\mathbb{F}$ then it has a solution in $\mathbb{Q}$.

Proof Let x_1 be a root in $\mathbb{F}$ then $x_1 \in \mathbb{F}_j$ for some j. If $j = 0$, we are done. Otherwise, let

$$x_1 = p + q\sqrt{r}$$

with $p, q, r \in \mathbb{F}_{j-1}$. We will show that there must be a solution in $\mathbb{F}_{j-1}$. We have

$$0 = (p + q\sqrt{r})^3 - m = p^3 + 3p^2 q\sqrt{r} + 3pq^2 r + q^3 r^{3/2} - m,$$

or grouping

$$0 = (p + q\sqrt{r})^3 - m = (p^3 + 3pq^2 r - m) + (3p^2 q + q^3 r)\sqrt{r}.$$

Let

$$\alpha = p^3 + 3pq^2 r - m, \quad \beta = 3p^2 q + q^3 r.$$

If β is non-zero then

$$\sqrt{r} = -\alpha/\beta$$

which shows that $p + q\sqrt{r} \in \mathbb{F}_{j-1}$. Otherwise,

$$\alpha = \beta = 0.$$

In this case,

$$(p - q\sqrt{r})^3 - m = \alpha - \beta\sqrt{r} = 0.$$

So $x_2 = p - q\sqrt{r}$ is also a root. We have found two real roots of our equation unless $q\sqrt{r} = 0$. If it does, then our original root was in $\mathbb{F}_{j-1}$ and we are done.[1]

We will now show that there if there are two real roots then there is a third and that it is in $\mathbb{F}_{j-1}$. We use the fact that a cubic real polynomial that has two real zeros has three real zeros. We prove this below. Call the roots x_j, we can write

$$x^3 - m = (x - x_1)(x - x_2)(x - x_3) = x^3 - (x_1 + x_2 + x_3)x^2 + \gamma x + \delta$$

for some real numbers γ and δ. We must have that the coefficient of x^2 is zero since it is zero in $x^3 - m$, so

$$x_3 = -x_1 - x_2 = -p - q\sqrt{r} - p + q\sqrt{r} = -2p.$$

Since p was in $\mathbb{F}_{j-1}$, we have a root in $\mathbb{F}_{j-1}$.

Repeating, we have a root in $\mathbb{F}_0 = \mathbb{Q}$ as claimed.   □

Now consider the case where m is not a perfect cube such as when $m = 2$. We know from Sect. 5.2 that the cube root is not rational, so there cannot be a root in $\mathbb{F}$. We thus have lots of examples of real numbers that are not in $\mathbb{F}$.

We used a lemma without proof.

Lemma 6.1 *If a cubic polynomial with real coefficients has 2 real roots then it has 3 real roots.*

There are multiple ways to prove this. First, we proceed in an elementary way.

Proof To see this divide the cubic polynomial, $p(x)$, by $(x - x_1)(x - x_2)$. We can then write

$$p(x) = (ax + b)(x - x_1)(x - x_2) + cx + d,$$

for some real numbers a, b, c, d. Putting $x = x_i$ for $i = 1, 2$ yields

$$cx_i + d = 0.$$

Since this is true for two distinct values of i, we must have $c = d = 0$. We then have that

$$x_3 = -a/b$$

is a root. So we have a third real root.   □

---

[1] In fact, $x^3 - m$ only has one real root and an alternate route to proving this theorem is to show that.

One alternate proof uses the fundamental theorem of algebra which we prove in Chap. 19. This implies that every cubic polynomial has three complex roots. It is then a question of showing that the third root is real. If the roots are z_j we know that z_1, z_2 are real. We also know that

$$z_1 + z_2 + z_3 = -a_2,$$

where a_2 is the coefficient of x^2 which we have assumed to be real. So $z_3 = -a_2 - z_1 - z_2$ is also real. Alternatively, it is easy to show that the complex conjugate of any complex zero of a real polynomial is also a zero. If z_3 is not real, its complex conjugate would yield a fourth root which is impossible so z_3 is real.

6.7 Problems

Exercise 6.1 Check that the intersection-enclosure pattern does indeed apply to each of the examples of Sect. 6.2.

Exercise 6.2 Given 3 points in the plane, what is their convex hull? Distinguish according to whether the 3 points are collinear.

Exercise 6.3 Given a subset of the plane, how many steps are required for the generation algorithm for convexity to terminate?

Exercise 6.4 A subset of $\mathbb{R}$ is said to be binarily division invariant if dividing any element by 2 results in an element of the subset. Check that intersection-enclosure applies. Also analyze generation for this property and show that it terminates with A_∞.

Exercise 6.5 A subset of $\mathbb{R}$ is said to be binarily division invariant and zero-happy if dividing any element by 2 results in an element of the subset, it is closed, and, in addition, if it contains zero then it also contains 2. Check that intersection-enclosure applies. Also analyze generation for this property and show that it does not always terminate with A_∞.

Exercise 6.6 A subset of $\mathbb{R}$ is said to be *open* if its complement is closed. Will there be a smallest open subset containing $[0, 1]$?

Exercise 6.7 Show that if a p is a polynomial with real coefficients and $p(z) = 0$ then $p(\bar{z})$ is also zero. Use this fact in conjunction with the fundamental theorem of algebra to show that every real polynomial of odd order has a real zero.

Chapter 7
Difference of Invariants

7.1 Introduction

One of the main themes in modern mathematics is the classification of types of objects. Given two objects of a given type, are they in some sense equivalent? Can we find a simple way to tell? One common to approach is to associate numbers to the objects in such a way that equivalent objects have the same numbers.

If two objects have distinct numbers then they are truly different. The related problem of finding a set of numbers that truly characterises a class of objects is generally much harder. Indeed, the characterising properties may not even be numbers but instead some simpler other class of objects.

7.2 Dominoes and Triminoes

We start with a simple example. Consider a chess-board from which two opposite corners have been removed. We have a set of dominoes. Each domino is the precise size of two squares of the chess-board. Is it possible to completely cover the board with dominoes in such a way that there is no overlap or protruding pieces? What about the same problem using triminoes? A trimino covers 3 squares of the board, and it may be straight or angular.

An alternate way of phrasing this problem is "is it possible to build a board of this shape by sticking dominoes (or triminoes) together?" We can generalize further by allowing the sticking of dominoes to the edge before they are laid on top. Indeed, we could view laying on top as cutting off a domino shape. The problem then becomes "what board shapes can be transformed into which other shapes?" Our original problem is now "can the board minus opposite corners be transformed to the empty board?"

In fact, the triminoes problem is easier than the dominoes one. There are 62 squares left on the board. A trimino covers 3 squares. The remainder when we divide 62 by 3

© Springer International Publishing Switzerland 2015

M. Joshi, *Proof Patterns*, DOI 10.1007/978-3-319-16250-8_7

is 2. Adding or subtracting a trimino will not change this number. We will therefore never get to zero. It is not possible to cover the board with triminoes and we are done. In this case, our invariant is the remainder on dividing the number of squares by 3. The board and the empty board have different invariants so we cannot go from one to the other. We have used *difference of invariants* to establish the impossibility.

Now suppose we remove the other two corners. The number of remaining squares is now 60. The remainder on dividing by 3 is 0 so they have the same invariant. It may therefore be possible to cover with triminoes. Or it may not! The difference of invariants only establishes impossibility. If the invariant is the same, it does not tell us much.

What about dominoes? The board has 62 squares which is a multiple of 2 so working with the remainder on division by two is not going to help. We therefore have to look for another invariant. A domino will always cover one black square and one white. We therefore use the number of black squares minus the number of white squares as our invariant. For a chess-board with all its corners, this invariant is zero. If we subtract two opposite black corners, we get -2, and for two opposite white corners, we get 2. The empty board clearly has zero. Subtracting a domino shape or adding one will not affect this invariant. We can therefore never get to zero. No matter how small we get, we will always be left with two more squares of one colour than the other. Note that in this case, if we subtract two adjacent corners, the invariant is zero, and it is, in fact, rather easy to cover the board with dominoes.

We have seen in a rather simple setting that the use of an invariant can establish that it is impossible to go from one object to another. The crucial point was that we found a number that did not change under each operation. This number was different for the two original objects and so we could not go from one to the other.

7.3 Dimension

All mathematicians and most users of mathematics learn linear algebra at some point in their career. A key invariant in linear algebra is *dimension*. In this section, we explore this concept and look at how to use it to establish that linear sub-spaces are different in a linear sense.

We shall say that a subset, E, of $\mathbb{R}^n$ is a vector sub-space if it is closed under addition and scalar multiplication. So

$$x, y \in E \implies x + y \in E,$$
$$x \in E, \lambda \in \mathbb{R} \implies \lambda x \in E.$$

In Sect. 6.4 a proof that every subset, A, of $\mathbb{R}^n$ is contained in a smallest sub-space, E, was given. The subset A is then said to generate E. In fact, the set E is given by all finite linear combinations of elements of A. So if

$$A = \{a_1, a_2, \ldots, a_d\}$$

then every element of E can be written as

$$e = \sum_{j=1}^{d} \lambda_j a_j$$

for some real numbers λ_j. To see that this holds, first observe that the set of all such vectors is a vector sub-space that contains A. So by the intersection-enclosure pattern it contains E. Second, any sub-space that contains A must contain all such sums by definition so E must contain them.

We shall say that a map,

$$T : U \to V,$$

between sub-spaces U and V is a *linear map*, if

$$T(\lambda x + \mu y) = \lambda T x + \mu T y, \tag{7.3.1}$$

for all $x, y \in U$ and $\lambda, \mu \in \mathbb{R}$. If there exists a linear map $S : V \to U$ such that ST is the identity map on U and TS is the identity on V then we say that U and V are *linearly equivalent*. Note that since T has a two-sided inverse, it is necessarily a bijection.

We shall say that a vector sub-space, V, is of dimension d if the smallest set that generates it is of size d. If a vector sub-space is of dimension d, then there are vectors $v_1, v_2, \ldots, v_d$ in V such that every element v of d can be written in the form

$$v = \sum_{j=1}^{d} \lambda_j v_j. \tag{7.3.2}$$

Such a minimal set of elements $\{v_k\}$ is called a *basis*.

The representation is unique: if we have

$$\sum_{j=1}^{d} \mu_j v_j = v = \sum_{j=1}^{d} \lambda_j v_j,$$

and $\lambda_k \neq \mu_k$ for some k, then

$$v_k = \frac{1}{\mu_k - \lambda_k} \sum_{j \neq k} (\lambda_j - \mu_j) v_j.$$

We can then use this equation to substitute for v_k and thus eliminate v_k from the right hand side of (7.3.2), and thus express every element v using $d - 1$ elements which contracts the minimality of d.

Consider $\mathbb{R}^3$. Let U_i be the sub-space of elements which may be non-zero only in coordinate i. So U_1 consists of vectors of the form

$$\begin{pmatrix} x \\ 0 \\ 0 \end{pmatrix}.$$

Let V_i be elements that are zero in coordinate i. So V_1 is the set of vectors of the form

$$\begin{pmatrix} 0 \\ y \\ z \end{pmatrix}.$$

The sub-spaces U_i are clearly of dimension 1. They can be generated by the vector with 1 in the appropriate slot, and they can clearly not be generated by 0 vectors!

The sub-spaces V_i are of dimension 2. Clearly, V_1 can be generated by

$$\begin{pmatrix} 0 \\ 1 \\ 0 \end{pmatrix}, \begin{pmatrix} 0 \\ 0 \\ 1 \end{pmatrix}.$$

We need to show that it cannot be generated by a single vector. If we take a fixed vector

$$\begin{pmatrix} 0 \\ x \\ y \end{pmatrix},$$

and y is zero then it clearly does not generate and we get U_2. If y is non-zero then we get vectors

$$\begin{pmatrix} 0 \\ \lambda x \\ \lambda y \end{pmatrix},$$

with λ an arbitrary element of $\mathbb{R}$. If we divide the second coordinate by the third then we get the same value, x/y, for every non-zero λ. This is clearly not the case for general elements of V_1 since it contains

$$\begin{pmatrix} 0 \\ 1 \\ 1 \end{pmatrix}, \begin{pmatrix} 0 \\ 0 \\ 1 \end{pmatrix}.$$

So V_1 cannot be generated by a single element and has dimension 2.

Note that the only difference between the definitions of V_1, V_2 and V_3 is the labeling of the entries. We can permute the entries to make each one into either of the others. We can therefore say that by *rearrangement,* the same result follows for V_2 and V_3.

We have seen that the sub-spaces U_i are of dimension 1 and the dimension of the sub-spaces V_j is 2. We now use this fact to show that they are not linearly equivalent. We do so by showing that if two sub-spaces are linearly equivalent then they are of the same dimension.

So suppose U and V are linearly equivalent. If U is of dimension d then there exists a basis

$$v_1, v_2, \ldots, v_d.$$

We also have a map $T : U \to V$ with inverse S. We show that

$$T v_1, T v_2, \ldots, T v_d$$

generates V. If $w \in V$ then $w = T(Sw)$. Now, by the definition of a basis, there exists scalars λ_j such that

$$Sw = \sum_{j=1}^{d} \lambda_j v_j$$

and so

$$w = T(Sw), \tag{7.3.3}$$

$$= T\left(\sum_{j=1}^{d} \lambda_j v_j\right), \tag{7.3.4}$$

$$= \sum_{j=1}^{d} \lambda_j T v_j, \tag{7.3.5}$$

where the final equality comes from the definition of a linear map. This shows that the vectors $\{T v_j\}$ generate V. The dimension of V is therefore at most d.

We also want to show that it is at least d. However, the same argument with U and V switched shows that the dimension of U is less than or equal to the dimension of V. We are done. Note that the last part of this argument was really another proof pattern. If two quantities are defined in identical manner and they have a symmetrical relationship, then if we can prove that one is less than or equal to the other, it follows that they are equal. The proof is simply to repeat the proof with the quantities switched. As a general rule, symmetry plus inequality implies equality.

It now immediately follows that the sub-spaces U_j are not linearly equivalent to the subspaces V_j since they are of different dimensions.

The case of linear subspaces is a rarity in that dimension is, in fact, enough to classify them. Two linear subspaces are linearly equivalent if and only if they have the same dimension. To see this, suppose U and V are of dimension d with bases

$$u_1, u_2, \ldots, u_d \text{ and } v_1, v_2, \ldots, v_d,$$

respectively.

Every element, u, of U can be written uniquely as

$$u = \sum_{j=1}^{d} \lambda_j u_j$$

as discussed above. We define

$$Tu = \sum_{j=1}^{d} \lambda_j v_j,$$

and similarly if

$$v = \sum_{j=1}^{d} \mu_j v_j,$$

we set

$$Sv = \sum_{j=1}^{d} \mu_j u_j.$$

It is immediate that ST and TS are both the identity map. Of course, we also need to show that T and S are linear. However, that is an easy exercise which we leave to the reader.

7.4 Cardinality

When we say two sets X and Y that have some structure are equivalent then we generally have a bijection between them. Both this bijection and its inverse are required to preserve the structure. In the sub-space example above, the requirement was linearity. In topology, they are required to be continuous.

A consequence is that whatever the extra structure required, there must exist a bijection between X and Y. Recall that a bijection is a map

$$f : X \to Y$$

such that every element y of Y is in the range of f, and precisely one element of X maps to it. So

$$\forall y \in Y, \exists x \in X \text{ such that } f(x) = y,$$

$$\forall x_1, x_2 \in X, f(x_1) = f(x_2) \implies x_1 = x_2.$$

Note that a bijection, f, automatically has an inverse g: we simply define $g(y)$ to be the unique element x such that $f(x) = y$. We can regard a bijection as a relabeling. We are simply giving the name $f(x)$ to the element x of X. In consequence, X and Y have the same number of elements.

In fact, for infinite sets, the existence of a bijection between two sets is typically taken as the definition of being the same size. They are then said to be of equal *cardinality*. A set is said to be *finite* if for some n it can be placed in a bijection with

$$\{0, 1, 2, \ldots, n - 1\}$$

and it is then said to have n elements. Note that all we are really doing is labeling the elements

$$x_0, x_1, x_2, \ldots, x_{n-1},$$

and so this is the same as our usual concept of a set having n elements. A set is *infinite* if it is not finite.

How can we use cardinality to prove that objects are not equivalent? We use commutative groups as an example. A commutative group is a set X with an operation $+$ with certain properties. There exists a special element, 0_X, and elements x, y, z in X, we have

$$x + y = y + x, \text{ commutativity,} \tag{7.4.1}$$
$$(x + y) + z = x + (y + z), \text{ associativity,} \tag{7.4.2}$$
$$x + 0_X = x, \text{ identity element,} \tag{7.4.3}$$
$$\forall x, \exists y, x + y = 0_X, \text{ inverse element.} \tag{7.4.4}$$

The first two properties say that $+$ is commutative and associative. The special element 0_X called the *identity element* which is commonly just called zero. Every element, x, has a negative, $-x$, that sums with it to zero. This negative is unique. If

$$x + y_1 = 0_X = x + y_2$$

then

$$y_1 = y_1 + (x + y_2) = (y_1 + x) + y_2 = y_2.$$

It is also possible to study groups that are not commutative. In that case, we generally write xy or $x * y$ rather than $x + y$. We will not study non-commutative groups here to retain simplicity.

Three simple examples of infinite commutative groups are $\mathbb{Z}$, $\mathbb{Q}$ and $\mathbb{R}$. However, there also finite examples. One standard example is the set of integers modulo k, denoted $\mathbb{Z}_k$. The underlying set is

$$\{0, 1, 2, \ldots, k - 1\}$$

which has k elements. We define $x + y$ in the usual way if $x + y$ is less than k. If it is not then we subtract k from it. This is sometimes called *addition modulo k*. The value 0 is clearly the identity and the inverse of x is $k - x$. Commutativity and associativity easily follow from the same properties of integers.

Given two commutative groups, we can easily construct another one by taking their Cartesian product and making addition act in each coordinate. If the groups are X and Y, we define
$$X \times Y = \{(x, y) : x \in X, y \in Y\},$$

with the operation

$$(x_1, y_1) + (x_2, y_2) = (x_1 + x_2, y_1 + y_2).$$

The element $(0_X, 0_Y)$ is the identity.

Two commutative groups X and Y are said to be *isomorphic* if there exists a bijection, f, which preserves $+$ so

$$f(x + y) = f(x) + f(y).$$

Note that for any $x \in X$,

$$f(x) = f(x + 0_X) = f(x) + f(0_X),$$

so $f(0_X) = 0_Y$. Similarly,

$$0_Y = f(0_X) = f(x + (-x)) = f(x) + f(-x)$$

for any $x \in X$. So
$$f(-x) = -f(x).$$

Since f is a bijection, it has an inverse f^{-1}. It turns out that f^{-1} is also an isomorphism. To see this,

$$\begin{aligned} f(f^{-1}(x) + f^{-1}(y)) &= f(f^{-1}(x) + f^{-1}(y)), \\ &= f(f^{-1}(x)) + f(f^{-1}(y)), \\ &= x + y. \end{aligned}$$

So on applying f^{-1} to both sides

$$f^{-1}(x) + f^{-1}(y) = f^{-1}(x + y).$$

In this particular case, f preserving the structure "+" and being a bijection was enough to make f^{-1} also preserve "+". The fact that a bijection preserves a given property implies its inverse also does so is quite a common phenomenon. However, it is by no means universal. For example, consider the function

$$f(x) = x^3$$

as a map from $\mathbb{R}$ to $\mathbb{R}$. The function f defines an infinitely differentiable bijection. Its inverse, g, is taking the cube root which is not differentiable at the origin. So for any differentiable function h, we have that

$$h \circ f$$

is differentiable. However, we need not have that $h \circ g$ is differentiable and in general, it will not be. The simplest example of failure is given by taking $h(y) = y$.

The solution generally adopted in such circumstances is simply to make a definition that the inverse must preserve the requisite property too! However, as we have seen, that is not necessary for commutative groups.

A great deal of effort has gone into classifying groups under isomorphism. How does cardinality help? Clearly, $\mathbb{Z}_k$ has k elements so it is not isomorphic to $\mathbb{Z}_r$ for any r not equal to k, nor is it isomorphic to $\mathbb{Z}$, $\mathbb{Q}$ or $\mathbb{R}$ since they are not finite.

In fact, one can prove that $\mathbb{R}$ has bigger cardinality than $\mathbb{Z}$ and $\mathbb{Q}$ and so cannot be isomorphic to either of them. See Sect. 13.4.

7.5 Order

We saw that cardinality is enough to show that $\mathbb{Z}_k$ and $\mathbb{Z}_r$ are not isomorphic for $r \neq k$. Now consider $\mathbb{Z}_k \times \mathbb{Z}_r$ and $\mathbb{Z}_{kr}$. Both of these groups have kr elements so cardinality is not going to help. If they are not isomorphic we need a different invariant.

Given an element x of a commutative group, we can define lx to be the result of summing l copies of x together. Using associativity, it does not matter how we do this addition:

$$lx = (l - 1)x + x = x + (l - 1)x.$$

We can then define the order of an element, x, to be the smallest positive natural number, m, such that

$$mx = 0.$$

If there is no such m, then x is said to be of infinite order. For integers, rationals, and reals all non-zero elements are of infinite order so the concept is not very useful. However, for finite groups, the situation is rather different. In particular, every element of a finite group has order at most equal to the group's cardinality. To see this suppose the group, X, has cardinality k, and $x \in X$. Consider the set of elements

$$X_{x,k} = \{x, 2x, 3x, \ldots, (k+1)x\}.$$

We know

$$X_{x,k} \subseteq X$$

and the cardinality of X is k. Two elements of $X_{x,k}$ must therefore be the same. (Pigeonhole principle!) We therefore have for some s, t, with $1 \leqslant s < t \leqslant k+1$

$$sx = tx,$$

which implies

$$(t - s)x = 0.$$

The value $m = t - s$ is at least 1 and at most k. We conclude that the order of x is at most k.

How does order help? If $f : X \to Y$ is an isomorphism, then we must have

$$f(0_X) = 0_Y$$

and

$$f(mx) = mf(x).$$

The same holds for the inverse of f. It follows that x and $f(x)$ have the same order. The set of orders of group elements must therefore be the same. If we can show two groups have different sets of orders then they are not isomorphic. Note the general point here, f is a bijection, both f and f^{-1} commute with $+$, so the two groups X and Y are the same except for the names of the elements. This means that any property defined in terms of "+" but not specific labels of the entries will be the same for both groups.

Consider $\mathbb{Z}_k \times \mathbb{Z}_r$ and $\mathbb{Z}_{rk}$. The order of 1 in $\mathbb{Z}_{rk}$ is rk simply by our definition of addition. What about $\mathbb{Z}_k \times \mathbb{Z}_r$? Some obvious elements are

$$(1, 0), (0, 1), (1, 1).$$

The first of these has order k and the second order r. The third will have order equal to the smallest m with $k | m$ and $r | m$ that is the least common multiple of k and r. This will equal kr if and only if k and r are co-prime.

We show that all elements of $\mathbb{Z}_k \times \mathbb{Z}_r$ have order less than or equal to this least common multiple, m.

$$m(x, y) = (mx, my) = (m.1.x, m.1.y) = (0.x, 0.y) = (0, 0).$$

We know $m.1 = 0$ in both $\mathbb{Z}_k$ and $\mathbb{Z}_r$ since r and k both divide m.

We have shown that the highest order of any element of $\mathbb{Z}_k \times \mathbb{Z}_r$ is the least common multiple of k and r, it is therefore not isomorphic to any group that has an element of order kr unless k and r are co-prime. Order has given us a way to distinguish commutative groups.

In fact, if two commutative groups X and Y have the same number of elements, m, and they both have an element of order m then they will be isomorphic. Let the two such elements be x and y. We define

$$f(lx) = ly$$

for $l = 0, 1, \ldots, m - 1$. Since x is of order m, the left-hand side ranges over all elements of X. (If two elements are the same, x is not of order m.) For the same reason, the right hand side ranges over all elements of Y and we have a bijection. Note

$$f(lx + rx) = f((l + r)x) = (l + r)y = lf(x) + rf(x),$$

provided $l + r \leqslant m - 1$. If $l + r > m$ then

$$f(lx + rx) = f((l + r - m)x) = (l + r - m)y = ly + ry = lf(x) + rf(x).$$

So f is indeed an isomorphism.

7.6 Divisibility

How can we show that $\mathbb{Z}$ and $\mathbb{Q}$ are not isomorphic? Both sets are infinite and it is possible to construct bijections between them so cardinality is not going to help. All non-zero elements are of infinite order so that will not help either. However, the fundamental difference is that elements of $\mathbb{Q}$ can be divided as many times as we like.

We can define the *divisibility order* of a non-zero element, x, of a commutative group X to be the largest positive integer m such that there exists y with

$$my = x.$$

If there is no largest value, we say that x has infinite divisibility order. For any $q \in \mathbb{Q}$, we can write

$$q = m\frac{q}{m}$$

for any positive integer m so all non-zero elements of $\mathbb{Q}$ have infinite order.

What about $\mathbb{Z}$? If $z \in \mathbb{Z}$, and $z > 0$, we can write

$$z = z1.$$

For $z < 0$, we have

$$z = (-z)(-1).$$

The order of a non-zero integer, z, is therefore $|z|$. Clearly, we cannot have a higher order since 1 is the smallest element and multiplying by a bigger number will yield too big an answer.

If two groups X and Y are isomorphic with isomorphism f then x and $f(x)$ must have the same divisibility order for all x in X. This is clearly not the case for $\mathbb{Z}$ and $\mathbb{Q}$ since all orders in $\mathbb{Q}$ are infinite and all ones in $\mathbb{Z}$ are finite. We have shown that $\mathbb{Z}$ and $\mathbb{Q}$ are not isomorphic by establishing that they have different sets of divisibility orders.

7.7 Problems

Exercise 7.1 What sizes of board $m \times n$, can be covered by 2×2 squares?

Exercise 7.2 Define multiplication on $\mathbb{Z}_k$ by taking xy and then taking its remainder on division by k. For what values of x and k does there exists a number y such that $xy = 1$?

Exercise 7.3 Let n be a positive integer. Let $f(n)$ denote the sum of its digits. Show that the remainder on division by 9 is invariant under passing from n to $f(n)$. Use this to show that if we keep summing up the digits of n until we get a number less than 10 then n is divisible by 9 if and only if this last number is 9. Repeat and reformulate this result for 3. More generally, does this approach work for any other numbers? What if we change our number base?

Chapter 8
Linear Dependence, Fields and Transcendence

8.1 Introduction

A key idea in linear algebra and many other areas of mathematics is that of dimension. The way in which dimension is defined varies according to context. In linear algebra it is is the number of vectors required to generate the entire space. We touched on this idea in Sect. 7.3. There we used it to show that linear maps could not be bijective between certain subsets of Euclidean space.

In this chapter, we explore a different application. We want to show that the set of *algebraic numbers* forms a field. A complex number, x, is said to be algebraic if there exists a non-zero polynomial p with rational coefficients such that $p(x) = 0$. Equivalently, one can require the coefficients to be integers. These two definitions are equivalent because one can multiply all the coefficients of a rational polynomial by the product of their denominators. Every rational number is algebraic. For $q \in \mathbb{Q}$ solves

$$x - q = 0.$$

Numbers that are not algebraic are said to be *transcendental*.

To show that the algebraic numbers form a field we have to show that if $p(x) = 0$ and $q(y) = 0$ for some rational polynomials p and q, then there are rational polynomials r and s such that

$$r(x + y) = 0, \text{ and } s(xy) = 0.$$

We also need a rational polynomial, t, such that

$$t(-x) = 0,$$

and if $y \neq 0$, we need another one, u, such that

$$u(y^{-1}) = 0.$$

© Springer International Publishing Switzerland 2015
M. Joshi, *Proof Patterns*, DOI 10.1007/978-3-319-16250-8_8

In fact, the last two are the easiest. If

$$p(x) = \sum_{j=0}^{N} c_j x^j,$$

for some c_j, then setting

$$d_j = (-1)^j c_j,$$

and taking these as the coefficients we have our polynomial, t, since

$$t(-x) = p(x) = 0.$$

For y^{-1}, we have

$$\sum_{j=0}^{M} e_j y^j = 0$$

for some $e_j \in \mathbb{Q}$ with $e_M \neq 0$. So multiplying by y^{-M}

$$\sum_{j=0}^{M} e_j y^{j-M} = 0$$

Putting $r = M - j$, we can rewrite this as

$$\sum_{r=0}^{M} e_{M-r} (y^{-1})^r = 0$$

and we have that y^{-1} is algebraic.

The other two cases are harder. Rather than trying to construct the appropriate polynomial explicitly, we proceed by using linear dependence.

8.2 Linear Dependence

We shall say that a subset, E, of $\mathbb{C}$ is *linearly dependent over* $\mathbb{Q}$ if there exists $e_1, \ldots, e_n \in E$ and $q_1, \ldots, q_n \in \mathbb{Q}$ not all zero such that

$$\sum_{i=1}^{n} e_i q_i = 0.$$

In words, we can take a rational amount of each e_j and sum to get zero. Note that this implies that for some j

$$e_j = -\sum_{i \neq j}(q_i/q_j)e_i.$$

In other words, at least one of the e_j can be written as a rational linear combination of the others.

If E is not linearly dependent, we shall say that it is *linearly independent*. We shall say that E *spans* a set $V \subset \mathbb{C}$ if any element of V can be written as a linear combination of elements of E with rational coefficients. So

$$v \in V \implies \exists q_j \in \mathbb{Q}, \text{ such that } v = \sum_{j=1}^{n} q_j e_j.$$

The crucial result we need is that if a subset F of V has more elements than a finite spanning subset E then F is linearly dependent.

We first shrink E to ensure that is linearly independent. If it is already linearly independent there is nothing to be done, otherwise some element, e_k, of E is a rational linear combination of other elements of E:

$$e_k = \sum_{j \neq k}\lambda_j e_j.$$

Now if $v \in V$ then for some $\{q_j\}$, we have

$$v = \sum_{j=1}^{n} q_j e_j = \sum_{j \neq k} q_j e_j + q_k \sum_{j \neq k}\lambda_j e_j$$

So we can discard e_k from E and still have a spanning set. We keep doing this until E is linearly independent and spanning. A linearly independent spanning set is called a *basis*. Note that if F had more elements than E before the discards, it certainly does afterwards so it is enough to consider the case that E is linearly independent.

We now suppose that E has n elements and is both linear independent and spanning. Consider F. Call its elements $\{f_j\}$. It has at least $n+1$ elements. If the first n elements are linearly dependent then so is F and we are done. It is therefore enough to consider the case that they are linearly independent. We will show that in this case they span V. This will be enough since $f_{n+1} \in V$ and so is in the set spanned by V which is equivalent to saying that it is a linear combination of the other elements. This says that F is linearly dependent.

We want to prove that the first n elements of F span if they are linearly independent. In fact, we will show that any subset of V that is linearly independent and has n elements is spanning. This is key to the concept of dimension in a linear context: all finite linearly independent spanning sets have the same number of elements. We can therefore say that the *dimension* of their span is equal to their size, and this definition makes sense since it is choice independent.

We proceed by induction. We will use permutations of $\{1, \ldots, n\}$. A permutation is a bijection on a set and can be simply thought of as a rearrangement of the numbers. Our inductive hypothesis is that there exists a permutation, σ, of $\{1, 2, \ldots, n\}$ such that if E_k is given by

$$E_k = \{f_1, \ldots f_k, e_{\sigma(k+1)}, \ldots, e_{\sigma(n)}\}$$

then E_k spans V. In other words, if we pick the right subset of E with $n - k$ elements then it together with the first k elements of F spans. If $k = 0$, the hypothesis is just that E spans which is certainly true. Now suppose the hypothesis holds for k and we want to prove it for $k + 1$. We have $f_{k+1} \in V$, so it is in the span of E_k and we can write

$$f_{k+1} = \sum_{j \leqslant k} \alpha_j f_j + \sum_{j \geqslant k+1} \beta_j e_{\sigma(j)}.$$

with $\alpha_j, \beta_j \in \mathbb{Q}$. The set F is linearly independent so f_{k+1} is not in the linear span of f_j with $j < k$. This means that we must have $\beta_j \neq 0$, for some $j > k$. We can relabel the elements of E so that $\beta_{k+1} \neq 0$. The rearrangement is equivalent to changing the permutation but that is within our inductive hypothesis and causes no problems. We use this new ordering to define E_{k+1} as the first $k + 1$ elements of F followed by the remaining elements of E post this permutation. Since $\beta_{k+1} \neq 0$, we have

$$e_{\sigma(k+1)} = \frac{1}{\beta_{k+1}} \left(f_{k+1} - \sum_{j \leqslant k} \alpha_j f_j - \sum_{j > k+1} \beta_j e_{\sigma(j)} \right).$$

To express any element, v, of V using E_{k+1}, we first write it as a linear combination of elements of E_k. We then substitute for $e_{\sigma(k+1)}$ using this expression. We have written it as a linear combination of elements of E_{k+1}. We have proven the inductive hypothesis for $k + 1$. It follows that the result holds when $k = n$, and the first n elements of F span.

The result we have just proven is sometimes called the *Steinitz exchange lemma*. It holds in much more general contexts than the one we have done here, but the proof is essentially the same.

Having proven that the first n elements of F span, it follows that f_{n+1} is a rational linear multiple of them. We conclude that F is linearly dependent as required.

8.3 Linear Dependence and Algebraic Numbers

We now show that the sum of two algebraic numbers is algebraic. First, observe that the statement that $x + y$ is algebraic is equivalent to the statement that there exists n such that the set

$$\{(x + y)^j, \ j = 0, \ldots, n\}$$

is linearly dependent over $\mathbb{Q}$. For both statements say that there exist rational numbers γ_j not all zero, such that

$$\sum_{j=0}^{n} \gamma_j (x + y)^j = 0.$$

Consider the set, V, of numbers spanned by numbers of the form

$$\{x^r y^s, \ r, s \in \mathbb{N}\}.$$

Whilst this set at first appears to be high dimensional, it is not so when x and y are algebraic. If they are algebraic, then there exists m, n such that

$$x^n = \sum_{j=0}^{n-1} q_j x^j$$

with q_j rational, and

$$y^m = \sum_{j=0}^{m-1} r_j y^j$$

with r_j rational. Using these expressions, we can therefore write

$$x^n y^m = \sum_{j=0}^{n-1} q_j x^j \sum_{k=0}^{m-1} r_k y^k$$

as a linear combination of lower powers and so it is in the span of

$$E_{m,n} = \{x^r y^s, \ r, s \in \mathbb{N}, r < n, y < m\}.$$

In fact, we can write x^l for any $l \geq n$ as a rational linear combination of x^j, $j < n$ by repeated substitution, and we can do similarly for y. This means that the set $E_{m,n}$ spans V.

All we have to do now is show that $(x + y)^l \in V$ for all l. As we showed above, since V is spanned by a set with mn elements, it cannot have a linearly independent subset with $mn + 1$ elements and it will follow that $x + y$ is algebraic. We have from the binomial theorem that

$$(x + y)^l = \sum_{i=0}^{l} \binom{l}{i} x^{l-i} y^i,$$

so it is indeed in V and we are done.

For xy, it is even easier: we have

$$(xy)^l = x^l y^l \in V.$$

So we must have that

$$\{1, xy, (xy)^2, \ldots, (xy)^{mn}\}$$

is linearly dependent, and we are done.

8.4 Square Roots and Algebraic Numbers

In Sect. 6.6, we saw the existence of a sub-field, $\mathbb{F}$, of $\mathbb{R}$ and therefore $\mathbb{C}$ which is closed under taking the square root of positive elements. We also saw that it was a proper subset of $\mathbb{R}$ that is there are real numbers that are not in $\mathbb{F}$. A natural question is how does $\mathbb{F}$ relate to the set of algebraic numbers? In fact, it is a proper subset.

First, we have previously shown that $2^{1/3}$ is not in $\mathbb{F}$ but it certainly solves

$$x^3 - 2 = 0$$

and so is algebraic. We conclude that the two sets do not coincide. Now observe that if $x > 0$ is algebraic so is $\sqrt{x}$ for if

$$p(x) = 0,$$

then

$$p((\sqrt{x})^2) = 0.$$

It therefore gives a zero of the polynomial q defined by

$$q(y) = p(y^2).$$

The set of real algebraic numbers is therefore a sub-field of $\mathbb{R}$ which is closed under the taking of positive square roots and it contains $\mathbb{Q}$. If we define $\mathbb{F}$ by intersection-enclosure, we have that the set of algebraic numbers is one of the sets that is intersected and so it must contain $\mathbb{F}$. In other words, $\mathbb{F}$ is a proper subset of the algebraic numbers as claimed.

Alternatively, we can use generation to show that $\mathbb{F}_j$ is a subset of the algebraic numbers for all j. To see this observe that every operation used to generate $\mathbb{F}_j$ from $\mathbb{F}_{j-1}$ does not cause us to leave the set of algebraic numbers.

8.5 Transcendental Numbers

We have shown that a lot of numbers are algebraic. Are all numbers algebraic? No. In fact, in a certain sense most numbers are not. We will prove this in Chap. 13. A simple example is a number in which we take an infinite decimal expansion and we insert vastly increasing numbers of zeros between successive 1 s. We let

$$w = \sum_{i=0}^{\infty} 10^{-i!}$$

This is called Liouville's constant. It was the first number that was proven to be transcendental. We sketch why this is true, a full proof is a little beyond our scope. Suppose we wish to show that w does not satisfy a polynomial of degree N. Its biggest coefficient will be less in magnitude than some power of 10 say 10^K. Consider the terms $w^0, w^1, \ldots, w^N$. Once we go far enough into the right tail of the expansion, the powers of the entries will be too far apart from each other to cancel and so we cannot get zero.

8.6 Problems

Exercise 8.1 Show that if $y \in \mathbb{C}$ is algebraic and $x^k = y$ then x is also algebraic.

Exercise 8.2 Show that if y is transcendental then y^k is transcendental for all counting numbers k. Show also that $y^{1/k}$ is transcendental.

Exercise 8.3 Show that the smallest sub-field of $\mathbb{R}$ which is closed under the taking of the powers $1/2$, $1/3$, $1/5$ is contained in the set of algebraic numbers.

Chapter 9
Formal Equivalence

9.1 Introduction

One of the most powerful tools in mathematics is to show that two seemingly unconnected classes of objects are in some sense equivalent. In other words, there is a map between the classes that is bijective and maps the essential operations of one class to those of the other. The more different the two classes are, the more powerful the technique is. A rather difficult to prove statement for one class may translate into a simple one for the other and vice versa. Sometimes we even have a map from a class to itself that transforms operations on the class.

Here we call this technique *formal equivalence*. We explore one example in depth, the equivalence of ruler and compass constructions with properties of sub-fields of $\mathbb{R}$. As a consequence, we show the impossibility of duplicating the cube.

9.2 Ruler and Compass Constructions

The ancient Greek mathematicians spent a great deal of effort on studying ruler and compass constructions. Here the objective is to construct a given angle or length using only a straight edge and a pair of drawing compasses. Note that "ruler" does not mean we can measure lengths. However, if we have two points a distance x apart then we can set the compasses to that size by fitting them to the two points. We can then draw a circle of radius x about any other point we have constructed. This also allows to measure the distance along a straight line from a given point on it.

In the original definition of a ruler and compass construction in Euclid's Elements, it was not possible to lift the compass off the paper without collapsing it. So one could draw a circle centred at a given point that passed through another given point but one could not copy a length from one place to another. However, the second Proposition in the Elements shows how to copy a length using a ruler and collapsible compass so allowing the use of a non-collapsing compass does not change the set of constructible

© Springer International Publishing Switzerland 2015
M. Joshi, *Proof Patterns*, DOI 10.1007/978-3-319-16250-8_9

points. For a beautiful presentation of the first six volumes of Euclid's Elements, see
the 2010 Taschen edition which is a colour reprint of an 1847 edition that replaced
names of lines with colours.

We generally assume that we are given two points a unit distance apart. We may
also be given some other lengths to take functions of. We can always draw a straight
line through any two points we have constructed and extend any straight line. A
typical problem is given a number x, find a given function of it. For example, the
ancient Greeks could construct the square root of two but not its cube root. Natural
questions which arise are

- can every number be constructed?
- how can we prove that certain numbers cannot be constructed?

Before proceeding to these we look at how to perform some basic constructions.

We first look at the construction of perpendiculars and parallels. To construct a
perpendicular to a given line through a point A, first mark off the two points that are
a distance 1 from A and call them B and C. We can do this by drawing a unit circle
centred at A and taking the points of intersection. Doing the same thing at B and C,
we get a further two points D and E which are at distance 2 from A. We now draw
a circle centred at B which passes through E and a second one centred at C that
passes through D. These two circles intersect in two points. Call one of these points
of intersection F. The line through A and F is perpendicular to the original line.
See Fig. 9.1. We can repeat the construction to get a line perpendicular to AF that

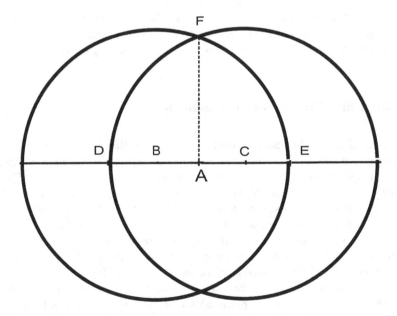

Fig. 9.1 How to construct a perpendicular to a given line using ruler and compass

passes through F. This second line is parallel to the original line through A since it is perpendicular to its perpendicular.

Whilst the last construction will construct a parallel line, often we want something better: we want a parallel line that goes through a given point. Suppose the line goes through points P and Q and we want a parallel line through a point A. Draw a circle centred at P through A. It will intersect with the line PQ twice. Take the point closer to A and call it S. Now draw two more circles of the same radius as the first centred at A and S. These two circles will intersect in two points: P and a new point R. The points $APSR$ define a rhombus since all of its sides are the same length. The sides of a rhombus are parallel so the line through A and R is parallel to our original line which goes through P and S. See Fig. 9.2.

If we have a length x and a length y, the first thing we might do is add them to get a length of $x + y$. Suppose we have a straight-line through a point A. We measure out the distance x from A along the line using our compass to get a point B. We then measure out the distance y from B to get a point C. The distance from A to C is now $x + y$. See Fig. 9.3.

Once we can add, the next question is how to subtract. We perform a similar construction. We measure x from A to get B. See Fig. 9.4. We then measure y back towards A to get C. Provided $x > y$, the distance from A to C will be $x - y$. We started with a unit length and we can add and subtract. We can now construct all the natural numbers by repeated addition.

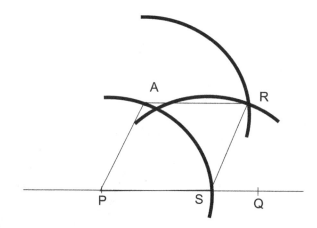

Fig. 9.2 How to construct a parallel line through a given point

Fig. 9.3 How to add two numbers using ruler and compass

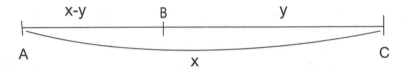

Fig. 9.4 How to take the difference of two numbers using ruler and compass

What other numbers can be constructed? The next obvious question is can we multiply and divide? The product and quotient of two lengths x and y can be constructed using similar triangles. We start with the lengths $1, x$ and y. First, we draw two intersecting straight lines. Call the intersection point A. On one of them mark off the distance 1 from A to a point B. We then mark off a further y from B to a point D. On the second line, mark off a distance x to a point C. We now draw a line from B to C. To finish, we draw a line parallel to the line BC through D. Note that this requires the construction of a parallel line that we did above. Denote by E its intersection with the line through AC. See Fig. 9.5. We claim that the distance from C to E is xy. To see this observe that the triangles ABC and ACE are similar since they have the same angles: they have the angle at A in common, and the other two angles arise from intersections of the same line with parallel lines so they must be the same. This means that we must have

$$\frac{|AB|}{|AD|} = \frac{|AC|}{|AE|}.$$

This says

$$\frac{1}{1+y} = \frac{x}{x+|CE|}.$$

Solving for $|CE|$, we find that it is equal to xy.

Fig. 9.5 How to construct the length xy from x and y

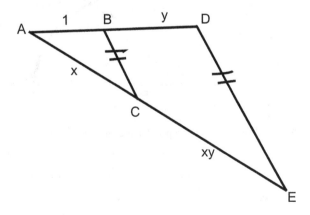

A related construction does division. We proceed as for multiplication but use different lengths. The length of AB is y, that of BD is 1 and that of AC is x. We have this time that

$$\frac{y}{1+y} = \frac{x}{x + |CE|}$$

from similarity. Solving yields that $|CE| = x/y$. See Fig. 9.6.

We start off with 1. We can add. We can divide. This means that we can construct any rational length. We can say that the set of positive rationals is constructible. Since we can multiply, we can also construct any rational multiple of a given number. Are there any other constructible numbers? Yes. For example, we can construct $\sqrt{2}$. Take a line and point on it. Construct a perpendicular through that point. See Fig. 9.7. Now measure a unit distance from the intersection point on both lines. It is immediate from Pythagoras's theorem that the distance between them is $\sqrt{2}$.

We can now construct the square root of any integer. We see how to construct $\sqrt{x+1}$ given $\sqrt{x}$. Mark the distance $\sqrt{x}$ along the perpendicular and 1 on the

Fig. 9.6 How to construct the length x/y from x and y

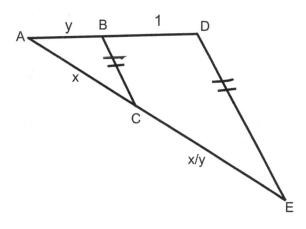

Fig. 9.7 How to construct $\sqrt{2}$ using ruler and compass

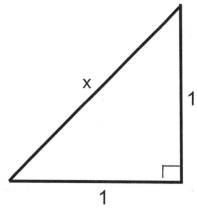

horizontal. The resulting right-angled triangle has hypotenuse of size $\sqrt{x+1}$ by Pythagoras's theorem. It follows by induction that the square root of any positive integer is constructible. Since the ratio of any two constructible numbers is constructible, so is the square root of any positive rational.

What about the square roots of square roots and their square roots as well? We now show that if a is constructible then so is $\sqrt{a}$. Assume that we have the length $a > 1$ and a unit length. We draw a and then 1 along a straight line with the point R in between. We take the midpoint of this line and call it P. We now draw a semi-circle with centre P of radius $(a+1)/2$. See Fig. 9.8. We also draw the vertical perpendicular to PR through R and denote its intersection with the semi-circle with Q. The length PQ is $(a+1)/2$ since Q is on a circle centred at P of that radius. The length PR is

$$t = \frac{a+1}{2} - 1 = \frac{a-1}{2}.$$

If s is the length QR, we know by Pythagoras's theorem that

$$s^2 + t^2 = \left(\frac{a+1}{2}\right)^2.$$

Hence,

$$s = \sqrt{\left(\frac{a+1}{2}\right)^2 - \left(\frac{a-1}{2}\right)^2} = \sqrt{a}.$$

There remains the square roots of positive numbers less than 1, however, these follow since

$$\sqrt{a} = \frac{1}{\sqrt{\frac{1}{a}}}$$

and $0 < a < 1$ if and only if $1/a > 1$.

Fig. 9.8 How to construct $\sqrt{a}$ using ruler and compass

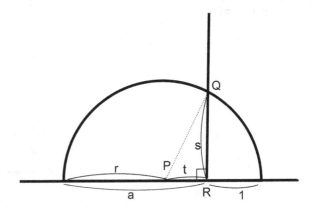

We have shown that the class of constructible numbers is closed under the operations of addition, subtraction, multiplication, division and square root taking. It also contains the positive integers. In fact, these properties characterize the class: it is the smallest subset of $\mathbb{R}$ that has these properties. How can we show this? The key is in observing how new lengths are constructed. We start with a unit length. We can pick a coordinate system so that its end points are $(0, 0)$ and $(1, 0)$. Every time we draw a line or make a circle, we have to use existing points. We can use them in three ways.

- We draw a line through two points.
- We measure the distance between two points for the radius of a circle.
- We use a point as the centre of a circle.

Once we have lines and circles, we use them by taking intersections. Again we have three possibilities.

- We intersect two lines and they have at most one point of intersection.
- A line and a circle intersect in zero, one or two points.
- Two circles also intersect in zero, one or two points.

Line equations are of the form

$$ax + by + c = 0$$

with a, b, c constants, and (x, y) varying over the line. Circles can be written

$$(x - \alpha)^2 + (y - \beta)^2 = \gamma^2,$$

with α, β, γ constants, and (x, y) varying over the circle.

Our equations of definition are quadratic or linear. When we solve for the solution set of two of them, we are either solving a linear equation, in which case we can get the solution from the equation's coefficients, purely using addition, subtraction, multiplication, and division, or we are solving a quadratic in which case we also have to use square roots. Crucially, these five operations are sufficient. Every solution point constructed must have coordinates arising from their repeated application. Any length is computable via the Pythagorean theorem which again only uses these.

We have shown that the set of constructible lengths, L, is equal to the positive elements of the smallest subset of $\mathbb{R}$ which contains the rationals, and is closed under the four basic arithmetic operations and taking square roots. We have replaced the problem of ruler and compass constructions with studying the closure of subsets of $\mathbb{R}$ under certain operations. This is an example of *equivalence*. A problem is replaced with one involving an apparently different class of object. In fact, we studied this equivalent problem in Sect. 6.6. There we showed that certain numbers are not generated by these operations. In particular, we showed that if $m \in \mathbb{Z}$ is not a perfect cube then $m^{1/3} \notin L$.

We can now solve the ancient Greek problem of duplicating the cube. Given a cube of side length x, is it possible to construct a cube with volume $2x^3$? This is equivalent to the question of constructing a cube with side length $2^{1/3}x$ which can be done if and only if we can construct the length $2^{1/3}$. We have shown that this is impossible and we are done.

Note that our proof shows that the construction is impossible if we are restricted to using straight edge and compass. We have said nothing regarding the possibility if other instruments are allowed. The impossibility of other ancient Greek problems has been shown in similar fashion. In particular, the trisection of an angle is impossible as is the construction of "squaring the circle". For this last, the problem is to construct a square with the same area as the unit circle. One therefore needs to be able a construct the length $\sqrt{\pi}$. This is constructible if and only if π is. However, π is *transcendental* which means that it is not a solution to any polynomial equation with integer coefficients. We showed in Chap. 8 that no constructible numbers are transcendental, and so π is not constructible. However, the proof of π's transcendence is a little too hard for this book.

9.3 Further Reading

Benjamin Bold's excellent book "Famous Problems of Geometry and How to Solve Them" is a very nice place to read more about the topics covered here.

9.4 Problems

Exercise 9.1 Using a ruler and compass construction, duplicate the square.

Exercise 9.2 Show directly that if a, b, r are rationals with $r > 0$, then $a + b\sqrt{r}$ satisfies a polynomial with rational coefficients.

Chapter 10
Equivalence Extension

10.1 Introduction

It is often the case that a class of objects with which we are working is lacking in some regard. For example, subtraction cannot be defined on the set of positive integers in a natural way. The solution is, of course, to allow negative numbers. Similarly, divisors do not always exist so we introduce fractions that is rationals to create them. Square roots of positive rationals are generally not rationals, so we introduce the reals. The square roots of negative reals do not exist in the reals either, so complex numbers are introduced.

Whilst we have a clear intuitive notion of integers and rationals, it is nevertheless an issue in mathematics how to construct them from simpler sets. Passing from the rationals to the reals is not even intuitively clear and they are many ways to do so. In this chapter, we study a common construction we call *equivalence-extension* which can be used to embed sets lacking a property into larger sets that possess it.

We review the concept of an equivalence relation in Appendix B, and here we look at how we can use them to construct sets.

10.2 Constructing the Integers

Suppose we have already constructed the natural numbers

$$\mathbb{N} = \{0, 1, 2, 3, \dots\},$$

and we have a natural notion of addition, $+$, on them. We now want to construct a bigger set on which every subtraction sum has an answer. For example, the problem *find x such that*

$$x + 2 = 1$$

has no solution in $\mathbb{N}$.

© Springer International Publishing Switzerland 2015
M. Joshi, *Proof Patterns*, DOI 10.1007/978-3-319-16250-8_10

Equivalence extension is one way to proceed. We first consider the set of all possible subtraction sums. This will just be the Cartesian product $\mathbb{N} \times \mathbb{N}$ since we are trying to subtract one number from another. Thus a pair (y, z) represents the sum $y-z$.

Even when sums have no answers in $\mathbb{N}$, we can often identify cases where they should have the same answer! For example, 2–3 and 4–5 should give the same result. We now define an equivalence relation on the set of sums so that they are equivalent if and only if they should have the same answers.

We *want* to say

$$(y_1, z_1) \sim (y_2, z_2)$$

if and only if

$$y_1 - z_1 = y_2 - z_2.$$

However, we cannot do so because the sums may not have solutions in $\mathbb{N}$ and we cannot use $\mathbb{Z}$, since that is what we are trying to construct! We therefore say

$$(y_1, z_1) \sim (y_2, z_2) \iff y_1 + z_2 = y_2 + z_1.$$

This is well-defined since addition is properly defined on $\mathbb{N}$. We need to check that $\sim$ is an equivalence relation.

First, $(y, z) \sim (y, z)$ for any y and z since

$$y + z = y + z.$$

So it is reflexive. For symmetry, we clearly have

$$y_1 + z_2 = y_2 + z_1 \iff y_2 + z_1 = y_1 + z_2,$$

and so

$$(y_1, z_1) \sim (y_2, z_2) \implies (y_2, z_2) \sim (y_1, z_1).$$

For transitivity, if $(y_1, z_1) \sim (y_2, z_2)$ and $(y_2, z_2) \sim (y_3, z_3)$, then $y_1 + z_2 = y_2 + z_1$ and thus

$$z_2 = (y_2 + z_1) - y_1$$

and this subtraction is well defined since it has an answer in $\mathbb{N}$. Similarly,

$$z_2 = (y_2 + z_3) - y_3.$$

We therefore have

$$(y_2 + z_1) - y_1 = (y_2 + z_3) - y_3.$$

Adding $y_1 + y_3$ to both sides yields

$$y_2 + z_1 + y_3 = y_2 + z_3 + y_1,$$

and so

$$z_1 + y_3 = z_3 + y_1.$$

This says

$$(y_1, z_1) \sim (y_3, z_3),$$

and $\sim$ is indeed transitive.

We have defined an equivalence relation on $\mathbb{N} \times \mathbb{N}$ and this induces a partition via its equivalence classes. What do the partition sets look like? The difference of the two elements is always the same, so we have upwards diagonally sloping lines. For example,

$$[(0, 0)] = \{(0, 0), (1, 1), (2, 2), (3, 3), \ldots\},$$

$$[(1, 0)] = \{(1, 0), (2, 1), (3, 2), (4, 3), \ldots\},$$

and

$$[(0, 1)] = \{(0, 1), (1, 2), (2, 3), (3, 4), \ldots\}.$$

Every equivalence class will have precisely one element that contains a 0. This will be $(n, 0)$ for $n > 0$, $(0, 0)$ or $(0, m)$ with $m > 0$. We therefore effectively have two copies of $\mathbb{N}$ joined together at 0. This is precisely what we want! The original positive natural numbers, zero, and our new numbers: the negatives of the counting numbers.

We still need to do a few things:

- define $+$ on the new set;
- show that $+$ extends the old $+$;
- show that every subtraction sum now has an answer.

We define

$$[(y_1, z_1)] + [(y_2, z_2)] = [(y_1 + y_2, z_1 + z_2)].$$

We have to be slightly careful; given an equivalence class it can be represented in more than one way, we have to make sure that our definition does not depend on the representation. This is a common issue in pure mathematics: we need to make sure that our definitions are well-defined when there is an arbitrary choice in a representation.

Thus we need to show that if

$$(y_1, z_1) \sim (\tilde{y}_1, \tilde{z}_1), \quad (y_2, z_2) \sim (\tilde{y}_2, \tilde{z}_2),$$

then

$$(y_1 + y_2, z_1 + z_2) \sim (\tilde{y}_1 + \tilde{y}_2, \tilde{z}_1 + \tilde{z}_2).$$

We have

$$y_1 + \tilde{z}_1 = \tilde{y}_1 + z_1,$$
$$y_2 + \tilde{z}_2 = \tilde{y}_2 + z_2.$$

Now

$$\tilde{y}_1 + \tilde{y}_2 + z_1 + z_2 = (\tilde{y}_1 + z_1) + (\tilde{y}_2 + z_2),$$
$$= (y_1 + \tilde{z}_1) + (y_2 + \tilde{z}_2),$$
$$= y_1 + y_2 + \tilde{z}_1 + \tilde{z}_2.$$

So the equivalence class of the sum is independent of the representation as required.

An alternate approach here would be to define $+$ in terms of distinguished elements. Each equivalence class has a unique member in which at least one entry is zero so we can define it in terms of that member. Thus we could define

$$[(x, 0)] + [(y, 0)] = [(x + y, 0)],$$
$$[(x, 0)] + [(0, y)] = [(x, y)],$$
$$[(0, x)] + [(0, y)] = [(0, x + y)].$$

We then have no troubles with the operation being well-defined. However, we would have to analyze special cases when studying the properties of operations which would be annoying. One solution is show that the two definitions agree, and then one simply uses whichever is convenient for a given proof thereafter.

The commutativity and associativity of "+" follow immediately from our original definition.

We call the set of equivalence classes $\mathbb{Z}$. We define an injection

$$i : \mathbb{N} \to \mathbb{Z},$$
$$i(n) = [(n, 0)].$$

This is clearly one-one since $m - 0 = n - 0$ if and only if $m = n$. It preserves "+" since

$$i(m + n) = [(m + n, 0)] = [(m, 0)] + [(n, 0)] = i(m) + i(n).$$

This means that we can identify $\mathbb{N}$ with its image under this map whilst retaining the same additive structure.

It remains to check that all subtractions now have an answer! Note

$$[(n, 0)] + [(0, n)] = [(n, n)] = [(0, 0)].$$

Since every element can be written with a zero in one of its two slots, this shows that every element, x, has a negative which we denote $-x$. For any x, y we can therefore always solve the equation

$$x + p = y$$

for an element p of $\mathbb{Z}$. We simply set

$$p = y + (-x).$$

We have used equivalence-extension to construct a set containing the natural numbers which is closed under subtraction, and corresponds to our intuitive notion of the integers.

10.3 Constructing the Rationals

The integers are great for addition, subtraction and multiplication, but not so good for division. We can, however, again apply the equivalence-extension pattern to create the rationals. Thus we consider the set of all division sums and identify sums that should have the same answer.

A division sum is a number to be divided and a non-zero number to divide by. This means that the set we need to partition is $\mathbb{Z} \times (\mathbb{Z} - \{0\})$ rather than $\mathbb{Z} \times \mathbb{Z}$. Our equivalence relation is

$$(p_1, q_1) \sim (p_2, q_2) \iff p_1 q_2 = p_2 q_1,$$

since the relation we want to express is $p_1/q_1 = p_2/q_2$.

We need to check that $\sim$ is indeed an equivalence relation. Reflexivity and symmetry are obvious. We prove transitivity in a very similar way as for the integers. If

$$(p_1, q_1) \sim (p_2, q_2) \text{ and } (p_2, q_2) \sim (p_3, q_3),$$

then

$$q_2 = (p_2 q_3)/p_3 \text{ and } q_2 = (p_2 q_1)/p_1.$$

So

$$(p_2 q_3)/p_3 = (p_2 q_1)/p_1.$$

This implies

$$p_1 p_2 q_3 = p_3 p_2 q_1.$$

If $p_2 \neq 0$ then $p_1 q_3 = p_3 q_1$ and $(p_1, q_1) \sim (p_3, q_3)$. If $p_2 = 0$, then $p_1 = p_3 = 0$ from the definition of $\sim$, and again $(p_1, q_1) \sim (p_3, q_3)$.

We have that $\sim$ is an equivalence relation. We call its set of equivalence classes the set of rational numbers, $\mathbb{Q}$. What do the equivalence classes look like? Since the ratio of the two entries is always the same, if we plot the classes on $\mathbb{Z} \times \mathbb{Z}$, each one will lie in a straight line through the origin.

It will sometimes be convenient to choose a special representative of each class. We can take this to be our usual concept of "simplest terms." If we multiply p and q by -1, we do not change $[(p, q)]$ so we can assume that $q > 0$. Also, if a number u divides both p and q, $[(p, q)]$ will equal $[(p/u, q/u)]$. This means that we can assume that p and q are coprime. Any other representative $(\tilde{p}, \tilde{q})$ will then have

$$\tilde{p} = vp, \ \tilde{q} = vq$$

for some q.

We now need to define addition and multiplication on $\mathbb{Q}$. We define

$$[(p_1, q_1)] * [(p_2, q_2)] = [(p_1 p_2, q_1 q_2)]$$

which corresponds to our intuitive notion of multiplying fractions. We need to test that "$*$" is well defined. In other words, we must show that

$$(p_1, q_1) \sim (\tilde{p}_1, \tilde{q}_1), \ (p_2, q_2) \sim (\tilde{p}_2, \tilde{q}_2) \implies (p_1 p_2, q_1 q_2) \sim (\tilde{p}_1 \tilde{p}_2, \tilde{q}_1 \tilde{q}_2).$$

However, given the left-hand-side equivalences, we have

$$p_1 p_2 \tilde{q}_1 \tilde{q}_2 = (p_1 \tilde{q}_1)(p_2 \tilde{q}_2) = (\tilde{p}_1 q_1)(\tilde{p}_2 q_2) = \tilde{p}_1 \tilde{p}_2 q_1 q_2,$$

and the result is clear.

What about addition? We work with our intuitive notion. We define

$$[(p_1, q)] + [(p_2, q)] = [(p_1 + p_2, q)].$$

We need to check

- that there is always a q such that both elements of $\mathbb{Q}$ can be represented in the form $[(s, q)]$,
- and that the choice of q does not affect the answer.

For the first, suppose we have $x = [(p_1, q_1)]$ and $y = [(p_2, q_2)]$ then we can represent via

$$x = [(p_1 q_2, q_1 q_2)], \ y = [(p_2 q_1, q_1 q_2)],$$

and it is clear that the representation exists.

To see that the choice of q does not matter, we show all choices agree with one particular choice, q', which is easy to work with. Given $x = (p_1, q_1)$ and $y = (p_2, q_2)$, we assume, as above that p_i is coprime to q_i and that $q_i > 0$ for $i = 1, 2$. We take q' to be the least common multiple of q_1 and q_2. For any other

q, we must have $q'|q$, since both q_1 and q_2 divide into q. We can therefore write $q = rq'$ for some r.

We then have to show the representation using rq' in the second slot, i.e. as denominator, agrees with that using q'. We have

$$x = [(s_1, q')], \quad y = [(s_2, q')],$$

for some $s_1, s_2 \in \mathbb{Z}$. This yields

$$x + y = [(s_1 + s_2, q')].$$

Our alternative representation is

$$[(rs_1, rq')] + [(rs_2, rq')] = [(rs_1 + rs_2, rq')] = [(r(s_1 + s_2), rq')].$$

However,

$$(s_1 + s_2, q') \sim (r(s_1 + s_2), rq'),$$

and we are done.

Note the general technique here, if we wish to prove equality of all elements of some set, we do not prove that any two arbitrary elements agree. Instead, we fix a particular element with nice properties and show that all other elements agree with it.

We have now constructed $\mathbb{Q}$ with addition and multiplication. There is a clear identity element:

$$[(1, 1)] = [(p, p)]$$

for any non-zero p. Every non-zero element has a multiplicative inverse, we just take

$$[(p, q)]^{-1} = [(q, p)].$$

So any division sum involving a non-zero divisor can be solved exactly.

We also have an additive identity

$$[(0, 1)] = [(0, q)]$$

for any non-zero q. Every element has a negative:

$$-[(p, q)] = [(-p, q)].$$

We have constructed the rationals.

One important property of the rationals that we have not yet discussed is its ordering. We define

$$[(p_1, q)] \leq [(p_2, q)] \iff p_1 \leq p_2$$

when $q > 0$. Multiplying p_1, p_2 and q by a positive constant will not affect the inequality so the ordering is well-defined. There are many more things one could check about the rationals such as associativity, commutativity, the distributive law, and that the operations $+$ and $*$ interact with $\leq$ in the way one might expect. We will not carry out the checks here but we encourage the reader to do them!

Our initial objective was to extend the integers so that all division sums have an answer. To complete this objective, we need to embed the integers into the rationals. We use the map

$$i : \mathbb{Z} \to \mathbb{Q}$$

defined by

$$i(z) = [(z, 1)].$$

In more usual language:

$$z = \frac{z}{1}.$$

The map i is one-one and it is easy to see that

$$i(x + y) = i(x) + i(y),$$

so we have indeed extended $\mathbb{Z}$.

10.4 The Inadequacy of the Rationals

We have seen how to construct the rationals and that they have many desirable properties. However, they still are lacking in certain regards. First, as we saw in Sect. 5.2, many natural equations lack solutions in $\mathbb{Q}$. For example, for any $k > 1$, the equation

$$x^k = 2,$$

has no rational solutions.

Second, many sequences that we might expect to converge do not. This is closely related to our first problem. For example, suppose we take a sequence $y_k = \frac{m_k}{10^k}$ where m_k is chosen to be the biggest integer such that

$$y_k^2 \leq 2.$$

The irrationality of root two guarantees that $y_k^2 < 2$ for all k. Since we defined m_k to be maximal, we must have

$$m_{l+1} \geq 10 m_l$$

and

$$y_{l+1} \geq y_l$$

for all l. We also have $y_l < 2$ for all l. The sequence (y_l) is therefore increasing and bounded.

It does not converge, however; the sequence (y_l) represents the decimal expansion of $\sqrt{2}$ to k decimal places. If it is to converge to anything it must be $\sqrt{2}$ which we know is not rational. Of course, we need to prove that the sequence does not converge to a rational.

Suppose it does converge to l then $l^2 \leq 2$ since $y_k^2 \leq 2$ for all k. If l is rational then $l^2 \neq 2$, so $l < 2$. Now

$$l + 10^{-j} \to l$$

as $j \to +\infty$. So

$$(l + 10^{-j})^2 \to l^2 < 2.$$

This means that if we take j sufficiently large,

$$(l + 10^{-j})^2 < 2.$$

If we now take the decimal expansion of l to j places, and call it $\tilde{l}$, we can only have decreased it, so

$$(\tilde{l} + 10^{-j})^2 \leq (l + 10^{-j})^2 < 2.$$

We see that y_j has not been properly defined since we can increase it whilst not taking its square above 2. We have proven by contradiction that y_k does not converge to a rational.

That increasing bounded sequences do not converge is therefore another defect of the rationals. A closely-related defect is the failure of the *least upper bound* property or *supremum* property. This states that any subset that is bounded above should have a smallest upper bound. For example, the sets

$$\{x \in \mathbb{Q} \mid x < 2\}, \text{ and } \{x \in \mathbb{Q} \mid x \leq 2\},$$

both have the least upper bound of 2. However, the set

$$E_2 = \{x \in \mathbb{Q} \mid x^2 < 2\} = \{x \in \mathbb{Q} \mid x^2 \leq 2\}$$

has no least upper bound. The natural candidate for such a least upper bound is $\sqrt{2}$ which, as we know, is not in $\mathbb{Q}$. We still need to prove that no rational upper bound is the least one, however. However, we can argue similarly to above. If b is a rational and $b^2 < 2$ then by taking $b_j = b + 10^{-j}$ for j sufficiently large, we obtain a value b_j with $b_j^2 < 2$ and $b_j > b$ so b is not an upper bound.

If b is a rational upper bound, it follows that $b^2 > 2$ and b is positive. Let $c_j = b - 10^{-j}$. We then have that $c_j < b$ for all j and for j sufficiently large $c_j^2 > 2$. So c_j is a positive rational number with $c_j^2 > 2$ and $c_j < b$. The number c_j will also be an upper bound, since squaring is an increasing operation. We thus have that b is not the least upper bound. Since b was an arbitrary upper bound, there is no least upper bound in the set of rationals.

10.5 Constructing the Reals

If we want to have least upper bounds, square roots and convergence of bounded increasing sequences then we need to construct a bigger set. This set is generally called the *real numbers*. We can proceed again by extension–equivalence.

Our base set is the set of bounded above subsets of rationals which we denote $B\mathbb{Q}$. Thus each element of $B\mathbb{Q}$ is a *set* of rational numbers which is bounded above. Some examples of elements are the sets

$$\{x \in \mathbb{Q} \mid x < 0\},$$
$$\{1\},$$
$$\{x \in \mathbb{Q} \mid x \leq 0\},$$
$$\{x \in \mathbb{Q} \mid x \leq 10\},$$
$$\{x \in \mathbb{Q} \mid x^2 < 2\}.$$

We want to identify subsets which should have the same least upper bound. The main problem is that they may not have a rational least upper bound and we are trying to construct the reals so using the real least upper bound is not allowed. However, if two subsets have the same least upper bound then they have the same set of upper bounds in the rationals and that is a statement that makes sense within the rationals.

We therefore say $E \sim F$ if and only if E and F have the same set of rational upper bounds. This is trivially an equivalence relation on $B\mathbb{Q}$. We denote the set of equivalence classes by $\mathbb{R}$. It will be convenient in what follows to have a unique representative of each equivalence class. We can then work with these without any worries regarding dependence on the choice of representative. For such an approach to work, the choice must be *natural* (or *canonical*) rather than arbitrary. Here we will show that each class has an element which contains all the others and we will use that.

There is an easy way to get a biggest element of an equivalence class of sets: just take the union of all of them. So given an equivalence class $[X]$ with X a bounded above subset of $\mathbb{Q}$, we define

$$U_X = \bigcup_{Z \sim X} Z.$$

In other words, we take all the members of $\mathbb{Q}$ which are in some member of the equivalence class. Note that $X \subseteq U_X$. Of course, we need to check that $U_X \in B\mathbb{Q}$ and that $U_X \sim X$. The second of these implies the first in any case.

Suppose y is an upper bound for X. It is then an upper bound for all Z in $[X]$ since they have the same upper bounds. It is therefore greater than all elements of U_X that is it is an upper bound for U_X. So the set of upper bounds for X is contained in the set of upper bounds for U_X. Since $X \subseteq U_X$, any upper bound for U_X is also one for X, and we have that the two sets of upper bounds are the same, and so $U_X \sim X$ as required.

In addition, it follows from transitivity that if $X \sim X_1$, then

$$U_X = U_{X_1}$$

since the unions are taken across the same sets, and so U_X depends only on $[X]$ rather than X. We can therefore write $U_{[X]}$.

What special properties does $U_{[X]}$ have? It is a bounded-above set of rationals and if $q \in U_{[X]}$ then $r \in U_{[X]}$ for all $r \in \mathbb{Q}$ with $r < q$. To see this, for such an r consider the set

$$V_r = U_{[X]} \cup \{r\}.$$

We claim that $V_r \sim U_{[X]}$. First V_r is bigger than $U_{[X]}$ so any upper bound for V_r is also an upper bound for $U_{[X]}$. Any upper bound for $U_{[X]}$ is greater than q which is greater than r so it is also an upper bound for V_r and we have $V_r \sim U_{[X]}$. Since $U_{[X]} \sim V_r$ and the definition of U_X requires that the union be taken over all sets which relate to U_X, we must have that r is in $U_{[X]}$.

The set U_X has an additional property: its complement in $\mathbb{Q}$ has no smallest element. We first show that that every element of the complement is an upper bound. Let $x \in \mathbb{Q} - U_{[X]}$. If it is not an upper bound then there must be an element $q \in U_{[X]}$ with $q > x$. But the argument above shows that then $x \in U_{[X]}$ and we have a contradiction.

The question is therefore whether the least upper bound if it exists must be in U_X. The answer is yes, because the single element set containing it would have the same upper bounds as U_X and so by the definition of U_X, we have $\{x\} \sim U_X$ and so $\{x\} \subset U_X$.

Definition 10.1 A bounded-above subset of rationals, V, with the properties that if $q \in V$, then $(-\infty, q) \subset V$, and such that $\mathbb{Q} - V$ has no smallest element is called a *cut*.

We have seen that every equivalence class can be represented by a cut.

In fact, there is a one-correspondence between the cuts and the equivalence classes; if two cuts are distinct then they are in different equivalence classes. To see this, suppose V_1 and V_2 are distinct cuts. Without loss of generality, suppose there exists $q_1 \in V_1 - V_2$, (swap indices otherwise). We must then have $q_1 > r$ for all r in V_2 since otherwise the definition of a cut would imply $q \in V_1$. It follows that $q_1 \in \mathbb{Q} - V_2$ and it is not the smallest element of $\mathbb{Q} - V_2$ from the definition of a cut. We therefore have an interval of rationals, (q_2, q_1), in $V_1 - V_2$. The sets V_1 and V_2 therefore have distinct sets of upper bounds since $(q_1 + q_2)/2$ is an upper bound for V_2 and not for V_1.

The upshot of all this is that we can work with cuts to define the real numbers, and in fact this is often taken as a starting point. It is known as the *Dedekind cut* construction of the reals. However, by starting with equivalence classes, we have seen that there is a natural motivation for the construction in terms of providing the least upper bound property.

Let $2^{\mathbb{Q}}$ denote the set of all subsets of the rationals which is sometimes called its power set. Every cut is an element of this power set. We can embed the rationals in the set of cuts by taking the map

$$i : \mathbb{Q} \to 2^{\mathbb{Q}},$$

$$i : q \mapsto (-\infty, q].$$

This is clearly injective and the image sets are certainly cuts. For the construction to be useful, this map must not be onto. In other words, there needs to be cuts that are not of this form. An easy example is the one studied above, take X to be the negative rationals together with $q \in \mathbb{Q}$ such that $q^2 < 2$.

There is a natural ordering on the space of cuts. We simply say $A \le B$ if and only if $A \subseteq B$. We want to show that we now get the least upper bound property. Suppose we have a collection of cuts

$$\{X_\alpha\}$$

with α ranging over some index set I. We can let

$$X = \bigcup_{\alpha \in I} X_\alpha,$$

that is we take the union of all the cuts. Clearly, if $y \in X$, and $q < y$ for some rational q, then $y \in X_\alpha$ for some α. Since X_α is a cut, $q \in X_\alpha$, and $q \in X$. So X has the cut property of being closed downwards.

Now suppose the collection of cuts is bounded above in the sense that there is a cut Y such that

$$X_\alpha \subseteq Y$$

for all α. Since Y is a cut there is a rational u such that all elements of Y are less than u. It follows that all elements of X_α and thus X are also bounded above by u.

We also need $\mathbb{Q} - X$ to have no smallest element. However, this may not hold. For example, if we set

$$Z_n = (-\infty, -1/n] \subset \mathbb{Q},$$

then, Z, the union over n is the set of negative rationals. Its complement has a clear smallest element: 0. However, there is an easy cure for both this case and in general. We take the cut, D, which represents the equivalence class containing the union.

Since the representative cut is bigger than all elements of the equivalence class, it is certainly at least as big as the union and therefore of all of its elements, X_α.

It remains to show that D is the least upper bound. That is we need to show that there is no cut, C, such that for all α

$$X_\alpha \subseteq C \subset D,$$

with the second inequality strict. Suppose C is a cut containing X_α for all α. First note that C must contain X by the definition of a union. Either, it is in the same equivalence class as X, in which case it is D since we have shown that there is one cut per class, or it is not. If it is not then it has a different set of upper bounds than X. Since it is bigger than X this means that there is an upper bound for X that is not an upper bound for C. Since X and D have the same upper bounds, D has an upper bound that C does not: D is smaller than C and we are done.

We have established that the set of cuts has the least upper bound property and we have constructed the real numbers.

We still need to establish that the reals have the properties that we would expect them to have. For example, we need to show that we can extend addition and multiplication to them in a natural way, and that the extensions interact with the ordering in a natural fashion. We do not carry out the extension here but it is not particularly hard. Once the reals have been constructed we simply use them as numbers and do not think about cuts anymore. We also regard the rationals as a subset of the reals.

10.6 Convergence of Monotone Sequences

The reals were introduced to cure the defect that bounded above sets of rationals did not have least upper bounds. We also saw that the rationals had a second related defect, namely that increasing bounded sequences sometimes fail to converge. We will see in this section that the reals do not have that defect. Let x_n be an increasing sequence in $\mathbb{R}$ and suppose $x_n \leq u$ for all n.

Let l be the least upper bound of the set $\{x_n\}$. Such an l exists since $\{x_n\}$ is bounded above by u and we have proven that there is always a real least upper bound. In fact, l is the limit of the sequence x_n. To see this note that if

$$y < l,$$

then y being less than the least upper bound is not an upper bound. So there exists N such that $X_N > y$. Since x_n is increasing we have

$$x_n \geq y, \text{ for all } n \geq N.$$

Thus given any $\epsilon > 0$, set $y = l - \epsilon/2$ and we have

$$l - \epsilon < x_n \leq l \text{ for all } n \geq N.$$

This is the definition of a limit and $x_n \to l$. See Chap. 19.

10.7 Existence of Square Roots

We now look at the problem of the existence of square roots for positive real numbers. Suppose $x > 0$ is real. Let

$$E_x = \{y \in \mathbb{R} : y^2 \leq x\}.$$

This set is non-empty since it contains 0 and is bounded above since all elements are less than the maximum of x and 1.

Let l be the least upper bound of E_x. This will be the square root of x. How do we prove this? If $l^2 > x$ then consider

$$l_n = l - \frac{1}{n}.$$

The sequence l_n converges to l and $l_n^2 \to l^2 > x$. For n sufficiently large we have

$$l_n^2 > x$$

and $l_n < l$. This means that l_n is an upper bound for E_x so l is not the least upper bound. We have a contradiction. We conclude

$$l^2 \leq x.$$

If $l^2 < x$, we can proceed similarly. Let $m_n = l + 1/n$, then for n sufficiently large $m_n^2 < x$ and so $m_n \in E_x$ and l is not an upper bound. We conclude that $l^2 = x$ as required.

10.8 Further Reading

Two nice books which are good introductions to the embedding of mathematics in set theory are Halmos's "Naive set theory" and Enderton's "Elements of set theory." Both of these books treat the subject matter of this matter in greater depth as well as introducing the axiomatic approach to set theory.

10.9 Problems

Exercise 10.1 Prove that if k is a positive integer and x is a positive real number then x has a kth root in the real numbers.

Exercise 10.2 Prove that every subset of reals that is bounded below has a greatest lower bound.

Exercise 10.3 Construct the positive rationals directly from the natural numbers, and then construct the negative rationals from them. Establish a bijection between the rationals constructed this way and the rationals constructed the original way. Ensure that the bijection is the identity on the natural numbers and that it commutes with multiplication and division.

Exercise 10.4 Give examples of subsets of the reals that do not have greatest lower bounds.

Exercise 10.5 Show that a monotone decreasing sequence of real numbers is bounded below if and only if it is convergent.

Chapter 11
Proof by Classification

11.1 Introduction

One of the major strands of modern pure mathematics is classification. A category
of objects is defined and the objective is then to make a list of all its members. Whilst
mathematicians tend to regard such an activity as worthwhile in its own right and as
an essential part of understanding the objects, it is natural for a non-mathematician to
ask what is the point? One answer is that once all the objects with a set of properties are
classified, it becomes very easy to prove theorems about them. Rather than working
with the definition, one simply works down the list.

Similarly, if one wants to know whether an object with a collection of properties
exists, one can consult classification lists and see whether any of the known objects
has them. If the list is complete, then failure to find is proof of non-existence.

In this chapter, we look at some simple examples of this powerful technique. We
develop the classification of all Pythagorean triples that is positive integer solutions to
Pythagoras's equation and use it to show non-existence of positive integer solutions
to the equation

$$x^4 + y^4 = z^2.$$

11.2 Co-prime Square

As a first application of proof by classification, we look at the problem of showing
that if a and b are co-prime and ab is a perfect square then a and b are also perfect
squares. We use a classification that we have already developed: the uniqueness of
prime decompositions. If $m = ab$ is a perfect square then $m = k^2$ with k an integer.
We can write uniquely

$$k = p_1^{\alpha_1} p_2^{\alpha_2} \cdots p_n^{\alpha_n}$$

© Springer International Publishing Switzerland 2015
M. Joshi, *Proof Patterns*, DOI 10.1007/978-3-319-16250-8_11

for a sequence of ascending primes p_j and positive integers α_j. We can therefore write

$$ab = k^2 = p_1^{2\alpha_1} p_2^{2\alpha_2} \ldots p_n^{2\alpha_n}.$$

Each of a and b has its own prime decomposition. However, since prime decompositions are unique, we can write

$$a = p_1^{2\alpha_1 - \beta_1} p_2^{2\alpha_2 - \beta_2} \ldots p_n^{2\alpha_n - \beta_n}, \tag{11.2.1}$$

$$b = p_1^{\beta_1} p_2^{\beta_2} \ldots p_n^{\beta_n} \tag{11.2.2}$$

for some non-negative integers β_j with $0 \leqslant \beta_j \leqslant 2\alpha_j$. We have not yet used the fact that a and b are co-prime. Note that if $0 < \beta_j < 2\alpha_j$ then p_j divides both a and b and so they are not co-prime. So we have for all j that $\beta_j = 0$ or $\beta_j = 2\alpha_j$.

In other words, the power of the prime p_j in the decompositions is either zero or $2\alpha_j$. Since every prime has an even power, this immediately implies that both of a and b are perfect squares. We are done.

To summarize,

Proposition 11.1 *If a and b are co-prime positive integers and ab is a perfect square then so is each of a and b.*

Whilst our proof of this result was not particularly elegant, the use of the classification in terms of primes meant that it was easy and straightforward.

11.3 Classifying Pythagorean Triples

Recall that the theorem of Pythagoras states that if a right-angled triangle has sides a, b and c, and c is the side opposite the right-angle then

$$a^2 + b^2 = c^2.$$

If we allow a, b and c to be real numbers then we choose any values of a and b and get a corresponding value for c by taking

$$\sqrt{a^2 + b^2}.$$

There is therefore an infinite number of solutions and they are easy to list. However, if we restrict ourselves to positive integer solutions, the classification problem becomes much more interesting. Such solutions certainly do exist:

$$3^2 + 4^2 = 5^2;$$
$$5^2 + 12^2 = 13^2.$$

Also, if we have one solution we can construct an infinite number simply by multiplying all three numbers by positive integers, that is for any $k \in \mathbb{N}$,

$$a^2 + b^2 = c^2 \implies (ka)^2 + (kb)^2 = (kc)^2.$$

The first piece of our classification is therefore to restrict our attention to cases where a, b and c have no common divisor knowing that once we have found all such triples, we can easily generate the rest.

In fact, we can do better.

Lemma 11.1 *If $a^2 + b^2 = c^2$ and a, b and c are positive integers with no common divisors then they are pairwise co-prime.*

Proof Suppose p is a prime or 1 and divides a and b. Clearly, p divides a^2 and b^2 so p divides $a^2 + b^2$. Thus p divides a, b and c so p is 1. We have shown that a and b are co-prime.

The cases where p divides c and one of a and b follow similarly. □

Since the numbers are pairwise co-prime, at most one is even. We show that if a and b are odd then c is not a perfect square. This will imply that exactly one of a and b is even and that c is odd.

If $a = 2k + 1$ and $b = 2l + 1$ then

$$a^2 + b^2 = 4k^2 + 4k + 1 + 4l^2 + 4l + 1 = 4(k^2 + l^2) + 4(k + l) + 2.$$

The square of an odd number is odd so $a^2 + b^2$ is not the square of an odd number. However, the square of any even number is divisible by 4 and $a^2 + b^2$ is clearly not a multiple of 4, so it is not a perfect square.

We therefore have that one of a and b is even and that c is odd. Relabelling a and b if necessary, we can say that a is odd and that b is even. We can now write

$$b^2 = c^2 - a^2 = (c - a)(c + a).$$

All of b, $c - a$ and $c + a$ are even, so we can write the integer equation

$$\left(\frac{b}{2}\right)^2 = \frac{c - a}{2} \frac{c + a}{2}.$$

We next want to show that $x = \frac{c-a}{2}$ and $y = \frac{c+a}{2}$ are co-prime. If p divides both of them then it divides $x + y$ and $y - x$, that is it divides c and a. Since c and a are co-prime, p must be 1. So x and y are co-prime as claimed.

Applying Proposition 11.1, we have that x and y are perfect squares. We can write $x = u^2$ and $y = v^2$. Note that u and v are co-prime and that $u < v$. We also have

$$\left(\frac{b}{2}\right)^2 = u^2 v^2.$$

Taking square roots, $b = 2uv$. Gathering, we have shown that there exist co-prime positive integers u and v with $u < v$ such that

$$a = v^2 - u^2, \tag{11.3.1}$$
$$b = 2uv, \tag{11.3.2}$$
$$c = u^2 + v^2. \tag{11.3.3}$$

Note that since a and c are odd, u and v must have the extra property that one is even and the other is odd. Our classification is almost complete. We still need to show:

- Different pairs (u, v) with these properties lead to different triples (a, b, c).
- That any triple (a, b, c) developed from these formulas, yields a Pythagorean triple with co-prime entries. (That is we need to check the reverse implication.)

The first is easy to check. We can recover v from a and c, via

$$v = \sqrt{(a + c)/2}$$

and u from

$$u = \sqrt{(c - a)/2}.$$

Since u and v can be found from the values of a and c, it must be that if you change u or v then the triple (a, b, c) must change too.

For the second, we really have to check two things: that you get a Pythagorean triple and that its entries are co-prime. We compute

$$(v^2 - u^2)^2 + (2uv)^2 = v^4 - 2u^2v^2 + u^4 + 4u^2v^2 = v^4 + 2u^2v^2 + u^4,$$

and this equals c^2. So we do indeed have a Pythagorean triple. Now let p be a 1 or a prime that divides a and c. Since one of u and v is even and the other odd, we have that c is odd so p cannot be 2. We have that p divides

$$v^2 - u^2 \quad \text{and} \quad u^2 + v^2$$

so p divides $2u^2$ and $2v^2$. Since p is not 2, it therefore divides u^2 and v^2. As p is a prime or 1, it therefore divides u and v. Since we assumed that u and v are co-prime, p must be 1. We have that a and c are co-prime, as needed. Note that this is sufficient, since any divisor of all three of a, b and c must certainly divide a and c.

Our classification is complete:

Theorem 11.1 *There is a bijection between the set of Pythagorean triples and positive integer triples (k, u, v) with u and v co-prime, $u + v$ odd and $u < v$ given by*

$$(k, u, v) \mapsto (k(v^2 - u^2), 2kuv, k(u^2 + v^2)).$$

This means that any time we want to prove a result about Pythagorean triples, we can prove it about triples of this form instead.

Now that we have the list, what are the actual triples? Here are a few:

$$(1, 1, 2) \mapsto (3, 4, 5);$$
$$(1, 1, 4) \mapsto (15, 8, 17);$$
$$(1, 2, 3) \mapsto (5, 12, 13);$$
$$(2, 1, 2) \mapsto (6, 8, 10).$$

Note that when we take $k = 1$ and $u = 1$, we have that $c - a = 2$, so we immediately see that there are infinitely many Pythagorean triples which have two side lengths differing by two.

11.4 The Non-existence of Pythagorean Fourth Powers

Having classified all positive integer solutions of the equation $a^2 + b^2 = c^2$, what about

$$a^4 + b^4 = c^4?$$

This is, of course, a special case of Fermat's Last Theorem and it has long been known to have no solutions. We can prove it as an application of the classification of Pythagorean triples. We use some other standard techniques.

First, we actually prove that

$$X^4 + Y^4 = Z^2$$

has no positive integer solutions, putting $X = a$, $Y = b$, and $Z = c^2$ then proves the desired result. Proving the stronger result turns out to be easier. This is not unusual with induction-style proofs in that the stronger inductive hypothesis makes it easier to establish results.

Second, we use the technique of *infinite descent* which was invented by Fermat to solve this problem. This is essentially the contrapositive of complete induction. We show that if a solution exists then there must be another smaller solution. This shows that if there is no solution with $X + Y + Z < k$, then there cannot be one with $X + Y + Z = k$. Applying complete induction, there is no solution for any k and we are done.

As before, we first reduce to the case where X and Y are co-prime. If $X = px$, $Y = py$ then

$$X^4 + Y^4 = p^4(x^4 + y^4).$$

So $p^4|Z^2$ which implies that we can write

$$Z = p^2 z$$

and we have

$$x^4 + y^4 = z^2.$$

So we need only consider the case that X and Y are co-prime. At least one is therefore odd and we assume that Y is.

Using the classification of Pythagorean triples, with $a = X^2$, $b = Y^2$ and $c = Z$, there exists u and v co-prime with $u < v$ and $u + v$ odd such that

$$X^2 = 2uv,$$
$$Y^2 = v^2 - u^2,$$
$$Z = v^2 + u^2.$$

We have

$$Y^2 + u^2 = v^2.$$

Since u and v are co-prime, we have another co-prime Pythagorean triple and since Y is odd, it follows that u is even and v is odd. Applying the classification again, there exist co-prime a and b with $a < b$, and $a + b$ odd such that

$$u = 2ab,$$
$$Y = a^2 - b^2,$$
$$v = a^2 + b^2.$$

So

$$X^2 = 2uv = 4ab(a^2 + b^2).$$

Hence

$$\left(\frac{X}{2}\right)^2 = ab(a^2 + b^2).$$

We have that a and b are co-prime. It follows that ab and $a^2 + b^2$ are co-prime. To see this, suppose p is 1 or prime and divides both. Since $p|ab$, it must divide a or b. It also must divide

$$a^2 + b^2 + 2ab = (a + b)^2,$$

and therefore $a + b$. It therefore divides both a and b so it must be 1.

Using Proposition 11.1, we have that both of ab and $a^2 + b^2$ are perfect squares. Since a and b are also co-prime, the same proposition shows that they are also perfect squares. Writing $a = \alpha^2$, $b = \beta^2$, $a^2 + b^2 = \gamma^2$, we have

$$\alpha^4 + \beta^4 = a^2 + b^2 = \gamma^2.$$

So we have another solution and its method of construction ensures that it is smaller than the first one. Our result follows by infinite descent and we are done.

11.5 Problems

Exercise 11.1 Find all Pythagorean triples with one side equal to 12.

Exercise 11.2 Suppose a, b, c, k are positive integers and

$$c^k = ab,$$

with a, b co-prime. Does it follow that a and b are the kth powers of positive integers?

Exercise 11.3 Suppose we want to find all rational Pythagorean triples, that is right-angled triangles with all sides rational. Can we classify these?

Exercise 11.4 Consider the map from Pythagorean triples to the length of the longest side as a natural number. Is this map surjective? Is it injective? What about for the other two sides?

Exercise 11.5 Suppose we allow triangles to have negative side lengths. What differences does it make to the classification of Pythagorean triples?

Exercise 11.6 How many Pythagorean triples have two sides differing by 1?

Chapter 12
Specific-generality

12.1 Introduction

When attempting to tackle a problem, a mathematician will often break it down into cases. Some cases will be easier than others to tackle. Often the process is simply random: after making additional hypotheses the result becomes easy and so the problem is broken into the case where they hold, and the case where they do not. This approach is called *case analysis.*

In this chapter, we look at a related technique that is in many ways more powerful which we call *specific generality.* With this approach we divide the problem into two cases A and B, and we show that if the result holds in case A then it also holds in case B. Equivalently, if it does not hold in case B then it does not in case A. The approach does not require nor imply the truth of the result in either case; instead, it says that we only need to study case A. Once that case has been dealt with in either a positive or negative sense we are done. The *specific* case A implies the general case of A and B.

We look at two applications of specific-generality in this chapter. The first is that we show that it is sufficient to prove Fermat's Last Theorem for prime exponents, and for the second we look at how to reduce the four-colour problem to a subset of maps.

12.2 Reducing the Fermat Theorem

Fermat's Last Theorem, proven by Andrew Wiles with assistance from Richard Taylor in 1994, states there are no positive integer solutions to the equation

$$x^n + y^n = z^n$$

© Springer International Publishing Switzerland 2015
M. Joshi, *Proof Patterns*, DOI 10.1007/978-3-319-16250-8_12

when $n > 2$. We saw in Sect. 11.3 that it is certainly not true when $n = 2$ and indeed we classified all the possible solutions. We also saw in Sect. 11.4 that when $n = 4$ it does hold.

We now show that if we can prove the theorem for all n prime bigger than 2 then it holds for all $n > 2$. We first show that any number bigger than 2 has either a prime factor bigger than 2 or is divisible by 4. Let $n > 2$. If it is prime, we are done. If not it is composite. It therefore has some prime factors. Either these are all 2 in which case it is a power of 2 and so must be divisible by 4, or one is not 2, in which case it has a prime factor which is not 2 and again we are done.

If we have a solution

$$x^n + y^n = z^n,$$

we can write $n = ql$ with q either a prime bigger than 2 or equal to 4. We now have

$$x^{ql} + y^{ql} = z^{ql}.$$

Let $X = x^l$, $Y = y^l$, $Z = z^l$, and then

$$X^q + Y^q = Z^q.$$

We have shown that if there is a solution for some $n > 2$, then there is also a solution for n either a prime bigger than 2 or 4. We have already shown that there are no solutions when $n = 4$. We conclude that q must be prime.

We have established that if the result holds for n prime then it holds in general. Note, however, that we have not established the prime case and so whilst we have reduced the remaining workload, we have not proven that it holds for any case other than 4.

12.3 The Four-Colour Theorem

The Four-Colour theorem states that if we have a map which divides the plane into countries, or regions, and we wish to colour it in such a way that no countries with a common border are the same colour, then we only need four colours of paint. A country is a connected subset of the plane with a nice boundary. This result has a long history and was eventually proven in 1976 by Kenneth Appel and Wolfgang Haken. Their proof was rather contentious, however, in that it involved breaking the problem down into a large number of special cases and then checking each one using a computer. A computer-free proof has not yet been found.

A feature of the theorem is that regions are only considered to be adjoining if they have a common edge and not if they only have a common vertex. In fact, it is easy to see that the result is false if any two regions with a common vertex have to be coloured differently: take a pie and divide it into N pieces; all the pieces have the centre of the pie as a vertex and they would all have to be different colours. So not only is the Four-Colour theorem false if we require all countries with a common vertex to be differently coloured, but so is the N-colour theorem for all N.

How can we apply specific-generality to the four-colour problem? A general map can have any number of countries meeting at a vertex but we shall show that it is sufficient to consider the case where a maximum of three countries do so. Such a map is said to be *cubic*.

Suppose we have proven the result for maps with only 3 countries meeting at each vertex, and suppose we have a general map, M, which does not have this property. We take each vertex of M and blow it up slightly, and make the interior of the bubble a new country. See Fig. 12.1. This map, M', has the property that if two countries share a common edge in M then their deformations in M' also do. In addition, each vertex of M' belongs to at most 3 countries. We therefore colour M' and shrink the introduced bubbles to get a colouring of M. Since the shrinking does not introduce any new adjacencies, the colouring of M is valid if that of M' is. In other words, if M' can be coloured with N colours so can M.

We have shown that if all maps with at most 3 countries meeting at each vertex can be coloured with 4 colours then so can all maps. As usual with specific generality, we have not actually shown that the result is true in any case, however.

We can further apply specific-generality to the Four-Colour problem to reduce to the case where all countries have at least 4 sides. To see this suppose all maps containing countries with at least 4 sides are colourable. We deduce the general case by induction. A map with less than 5 countries is certainly 4 colourable. Suppose all maps with n countries are colourable. Take a map with $n + 1$ countries. If it has no countries with less than 4 sides, it is colourable by our general hypothesis. If it has countries with two or three sides, then take one such country and merge it with a neighbour. The resulting map has n countries so it is colourable. Colour it. The country that was merged has at most 3 neighbours so we can simply give it a colour different from them and we have coloured the map with $n + 1$ countries. Our general result follows by induction.

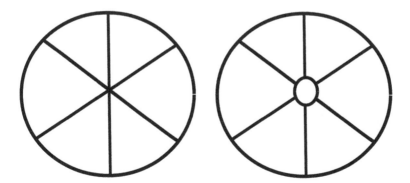

Fig. 12.1 A pie with six pieces: before and after the central vertex is made into an extra country

12.4 Problems

Exercise 12.1 Suppose we can prove that every map in which all countries have at least k sides can be coloured with k colours. Show that every map can then be coloured with k colours.

Exercise 12.2 Show that if there are no solutions to the Fermat equation with x, y, z pairwise co-prime then there are no solutions in general.

Chapter 13
Diagonal Tricks and Cardinality

13.1 Introduction

For non-mathematicians, the concept of infinity is rather nebulous and generally just refers to any number that is not finite. For mathematicians, there are numerous types of infinities depending on the nature of the object being studied. For example, we may wish to study how many elements are in sets and then we are studying *cardinality*. Alternatively, we may want a concept of a limit for divergent sequences—we often say that a sequence or series tends to infinity, but what does that mean? Finite numbers are often used to order objects as well to count them, we may wish to extend these concepts of ordering to sets that are not finite. Such numbers are called *ordinals*. The fact that the sets of ordinals and cardinals are the same for finite numbers does not imply that the concepts will be the same in the infinite case, and, in fact, they are different.

In this chapter, we will examine sizes of infinity from the point of view of cardinality. In particular, we will see that there are an infinite number of different sizes of infinity. We will also see that there are in a certain sense of the word *more* real numbers than rationals but the same number of rationals and integers.

13.2 Definitions

First, we define when two sets are the same size.

Definition 13.1 Two sets A and B have the same cardinality if there exists a bijection between them.

For finite sets, this corresponds to the notion that a set has n elements if and only if its elements can be labeled with the numbers 1 to n. Sometimes, it is convenient to label them $0, 1, 2, \ldots, n - 1$.

We can also say that a set has fewer elements than another one.

© Springer International Publishing Switzerland 2015
M. Joshi, *Proof Patterns*, DOI 10.1007/978-3-319-16250-8_13

Definition 13.2 A set A is of smaller cardinality than a set B if there exists an injective map from A to B but there does not exist a bijection between A and B.

For this definition, to be useful, we need to show

Theorem 13.1 (Schröder–Bernstein theorem) *If there exists an injective map from A to B, and one from B to A, then there exists a bijection between A and B.*

13.3 Infinite Sets of the Same Size

Many infinite sets are the same size. Quite unlike finite sets, we can find infinite sets which are the same size as some of their subsets. In fact, a set having the same cardinality as a proper subset is sometimes taken as a definition of its non finiteness. For example, consider the set of even natural numbers, E. In fact, E and $\mathbb{N}$ have the same cardinality. Let

$$f : \mathbb{N} \to E,$$

be defined by $f(n) = 2n$. This function is a bijection and so they have the same cardinality.

Similarly, the function $g(n) = n^2$ shows that there are the same number of perfect squares as natural numbers. In fact, any infinite subset, S, of $\mathbb{N}$ will have the same cardinality as it. To see this observe that it is naturally ordered since the natural numbers are. So we define a map, h, from $\mathbb{N}$ to S as follows. Let $h(0)$ be the smallest element of S. Let $h(1)$ be the smallest element of $S - \{h(0)\}$. We keep going and let $h(n)$ be the smallest element of $S - \{h(0), h(1), \ldots, h(n-1)\}$. The map h is certainly injective since each number mapped to is bigger than the previous one. After n steps, we will have covered all numbers in S that are less than n, so h is onto.

So any infinite subset of $\mathbb{N}$ is the same size as $\mathbb{N}$. In particular, there is the same amount of prime numbers as natural numbers. This is despite the fact that they get further and further apart along the number line.

Note also that we can always add a finite number of elements to a countable set and still have a countable set. To see this, suppose we want to add $X = \{x_1, \ldots, x_n\}$ to the natural numbers. Let $f(j) = x_{j+1}$ for $j < n$ and let $f(j) = j - n$ for $j \geq n$. We then have a bijection from $\mathbb{N}$ to $X \cup \mathbb{N}$.

What about larger sets? Now consider $\mathbb{N} \times \mathbb{N}$. This is the set of ordered pairs (m, n) with m, n both natural numbers. Is this any bigger? The answer is no. We can construct a bijection from $\mathbb{N}$ to it by enumerating along the diagonals. So

$$f(0) = (0, 0), \quad f(1) = (1, 0), \quad f(2) = (0, 1),$$

and

$$f(3) = (2, 0), \quad f(4) = (1, 1), \quad f(5) = (0, 2).$$

Every pair lies in some diagonal, and we go down these one by one. This means that every pair is in the image of f and we have a bijection. For future use, denote its inverse by g. Note that this result shows not just that $\mathbb{N} \times \mathbb{N}$ is countable but also that the Cartesian product of any two countable sets is countable. We simply label their elements via their bijection with $\mathbb{N}$ and the result is immediate.

We can use this result to establish the cardinality of $\mathbb{Q}$. We work with the non-negative rationals, $\mathbb{Q}_+$. We have the natural injection

$$i : \mathbb{N} \to \mathbb{Q}_+,$$

which is simply mapping each natural number to itself. We also have a simple map,

$$j : \mathbb{Q}_+ \to \mathbb{N} \times \mathbb{N},$$

which is to map a rational q to its numerator, m, and denominator, n, in lowest terms. So

$$j(m/n) = (m, n).$$

The map $g \circ j$ is then an injection from $\mathbb{Q}_+$ to $\mathbb{N}$. We thus have injections both ways when considering $\mathbb{Q}_+$ and $\mathbb{N}$. It follows from the Schröder–Bernstein theorem that they are of the same cardinality.

We also have that any countable union of countable sets is countable. Suppose E_j is countable for $j = 0, 1, 2, \ldots$. We can assume E_j is non-empty since if it is empty then it has no effect on the union. We then have a bijection f_j from E_j to either $\mathbb{N}$ or a subset of it. We can define

$$f : \bigcup E_j \to \mathbb{N} \times \mathbb{N},$$

by

$$f(x) = (j, f_j(x)) \text{ for } x \in E_j.$$

If x is in more than one E_j, use the smallest value of j for which $x \in E_j$. This yields an injection from $\bigcup E_j$ to $\mathbb{N} \times \mathbb{N}$ and so an injection to $\mathbb{N}$ since $\mathbb{N} \times \mathbb{N}$ is countable. If $\bigcup E_j$ is finite our result is trivial. If for some j, E_j is infinite, then we have a bijection from it to $\mathbb{N}$ whose inverse yields an injection from $\mathbb{N}$ to $\bigcup E_j$. Otherwise, in the case that all the sets are finite, we can construct an injection from $\mathbb{N}$ by simply enumerating all the elements of each set one by one, and skipping over any already in the image. It then follows from the Schröder–Bernstein theorem that since we have an injection each way that there is a bijection between $\bigcup E_j$ and $\mathbb{N}$. So $\bigcup E_j$ is countable.

13.4 Diagonals

After the above examples, the reader might be forgiven for thinking that all infinite sets are the same size. How can we prove that there are more real numbers than natural numbers? We show that any countable list of reals is incomplete by using a pattern sometimes called the *diagonal trick*. We employ decimal representations. So each real number on the list is written first as a decimal. In order to ensure that representations are unique, we make the convention that decimals cannot end in an infinite string of recurring 9s. So if we have a number

$$1.9999999999999999999999999999999\ldots$$

we write
$$2.0000000000000000000000000000000\ldots$$

instead.

The reader may not be familiar with the fact that a number ending in an infinite string of 9s can be represented by a finite string instead. We digress a little to discuss this fact. The key here is to realize that a decimal expansion is really just an infinite sum

$$\sum_{j=0}^{\infty} a_j 10^{m-j}$$

where $m \in \mathbb{Z}$ determines the magnitude of the number, and a_j are the digits lying in the range 0 to 9 with $a_0 \neq 0$. We will study infinite sums in Chap. 19. The sum is increasing and bounded by 10^{m+1} so it certainly converges. Suppose a number, x, has an expansion ending in an infinite string of 9's consider the last digit that is not 9 and suppose this is in place k (We assume that there is at least one non 9, the reader can deal with the case of all 9s.). Define a new expansion $b_i = a_i$ for $i < k$ and $b_k = a_k + 1$. We take zero thereafter. Intuitively, we have rounded up the trailing 9s. The implied real number, y, has a finite decimal representation.

Now consider $y - x$. The terms before k agree so after canceling we have essentially that $y - x$ equals
$$z = 1 - 0.999999\ldots$$

times some integer power of 10. Now z must be smaller than $1 - 0.9 = 0.1$ and $1 - 0.99 = 0.01$. So take a finite number of 9s, r, and consider that

$$z < 1 - \sum_{i=1}^{r} 9 \times 10^{-i} = 10^{-r-1}.$$

Since this is true for any r, z must be zero.

Back to studying the cardinality of $\mathbb{R}$, it is easier to work with the interval $(0, 1)$ which is certainly has fewer elements than $\mathbb{R}$. Let $f : \mathbb{N} \to (0, 1)$ be any map. We show that f is not surjective. We do this by constructing a number not in the image of f. We do this by making the number disagree in the nth decimal place with $f(n)$.

So let x_n be the digit in the n decimal place of $f(n)$. Let

$$y_n = (x_n + 2) \mod 10.$$

The notation mod means take the remainder on division by 10. In this case, it means that if the answer is 10 take 0, if it is eleven take 1.

Now let

$$y = \sum_{n=1}^{\infty} y_n 10^{-n}.$$

The number y has y_n in the nth decimal place unless $\{y_n\}$ ends in an infinite string of 9s in which case y is all zeros after some point. The nth decimal digit therefore differs from x_n since either x_n is 2 different from y_n, or x_n is 7 in which case it is still different from 0.

So y and $f(n)$ are different in the nth decimal place. This implies that

$$y \neq f(n).$$

So f is not surjective as claimed.

Why is this technique called the diagonal trick? If we write the numbers in vertical list, we are creating a new number from the digits on the diagonal. For example, if we have for the first 9 numbers

0.693458302,

0.780695725,

0.132661746,

0.723736632,

0.746140245,

0.969628025,

0.948094516,

0.949140569,

0.477748785,

then the diagonal digits are

6, 8, 2, 7, 4, 8, 5, 6, 5.

The value of y is then

$$0.804960787\ldots$$

We have shown that no bijections from $\mathbb{N}$ to $\mathbb{R}$ exist. The set $\mathbb{R}$ is said to be *uncountable*. Note that injections from $\mathbb{N}$ to $(0, 1)$ certainly exist, however. We can set

$$g(n) = \frac{1}{n+1}$$

to get such an injection.

How can we construct even bigger sets? The standard way to do it is to take the set of all subsets. Thus if X is a set, let 2^X denote the set of its subsets. Note that each element of 2^X is a subset of X rather than an element of X. Thus if $X = \{1, 2, 3\}$ then

$$2^X = \{\varnothing, \{1\}, \{2\}, \{3\}, \{2, 3\}, \{1, 3\}, \{1, 2\}, \{1, 2, 3\}\}.$$

In this case, 2^X has $8 = 2^3$ elements. One of these elements is the empty set itself.

In general, if a finite subset has n elements then its set of all subsets has 2^n elements. Hence the notation. To see this observe that we can identify a subset of $\{1, 2, \ldots, n\}$ with a binary number. We put 1 in slot j if j is in the subset and 0 otherwise. We thus get a bijection from the set of subsets to the set of n digit binary numbers that is the set of numbers from 0 to $2^n - 1$. So the set of subsets has 2^n elements.

For infinite sets, of course, life is rather more complicated. However, we can use a variant of the diagonal trick to show that 2^X is bigger than X. Let

$$f : X \to 2^X.$$

We need to show that it is not surjective. We need to find a subset that is not in the image of f. Let

$$A = \{x \mid x \notin f(x)\}.$$

We now show that the set A cannot be in the image of f. Suppose for some y, $A = f(y)$. Either $y \in A$, or $y \notin A$. We show that both of these are impossible. If $y \in A$, then by the definition of A we have $y \notin f(y)$ but $f(y)$ is A so $y \notin A$ and we have a contradiction. Now for the other case, if $y \notin A$ then $y \notin f(y)$ so $y \in A$. Again, we have a contradiction.

There are no surjections from X to 2^X. However, there are certainly injections. We simply define

$$f(x) = \{x\}.$$

So 2^X has a bigger cardinality than X.

Starting with $\mathbb{N}$, we can repeat many times. Let $X_0 = \mathbb{N}$. Let $X_j = 2^{X_{j-1}}$. In each case, X_j has bigger cardinality than X_{j-1}. The cardinality of X_j is called $\aleph_j$. The Hebrew letter $\aleph$ is pronounced aleph.

We now have an infinite number of sizes of infinity! In fact, we have $\aleph_0$ different sizes. The next natural question is "are there any more?" The answer is "yes". If we take a set Y which is the union of all these sets (the fact that we can do this is a little deep for this book) it has more elements than X_j for every j and gives us a new order of infinity. We can then start taking 2^Y and so on. This sort of process will go on forever. There are an awful lot of sizes of infinity.

How big are the reals? We have seen that their cardinality is not $\aleph_0$. In fact, it is $\aleph_1$. There are a number of ways to see this. One is simply to observe that every subset of the natural numbers can be viewed as a binary number between zero and one. We do this by putting 1 in the jth place after the point if j is in the subset and zero otherwise. We thus get a bijection from $2^{\mathbb{N}}$ to $(0, 1]$. The sets $\mathbb{R}$ and $(0, 1)$ have the same cardinality. For example, the map

$$x \mapsto -1 + \frac{1}{x}$$

defines a bijection from $(0, 1)$ to $(0, \infty)$. The map

$$y \mapsto \log y$$

defines a bijection from $(0, \infty)$ to $\mathbb{R}$. So $(0, 1)$ and $\mathbb{R}$ have the same cardinality.

How do we show that $(0, 1)$ and $(0, 1]$ have the same cardinality? We define

$$f : (0, 1) \to (0, 1]$$

to be the identity on the irrationals. For the rationals, the intersection of $(0, 1)$ with $\mathbb{Q}$ is countable so, as we saw above, there is a bijection to its union with $\{1\}$. We use this bijection to define f on the rationals and we are done.

13.5 Transcendentals

We can use countability to prove the existence of transcendental numbers. We will prove that the set of algebraic numbers is countable. So, in fact, almost all real and complex numbers are transcendental. First, we show that the set of polynomials with rational coefficients is countable. Such a polynomial is just a finite sequence of rational numbers. The question therefore is equivalent to showing that the set of finite sequences of elements of a countable set is countable.

The set of polynomials of a given degree, d, is simply the set of rational sequences of length d. In other words, it is the set

$$\mathbb{Q}^d = \mathbb{Q} \times \mathbb{Q} \times \mathbb{Q} \ldots \mathbb{Q}$$

with d terms. We showed above that the Cartesian product of any two countable sets is countable. It then follows by induction that $\mathbb{Q}, \mathbb{Q}^2, \ldots, \mathbb{Q}^d$ are countable since $\mathbb{Q}$ is. The set of rational polynomials of degree d is therefore countable. A countable union of countable sets is rational so it follows that the set of all rational polynomials is countable.

A rational polynomial only has a finite number of zeros. If we take the zeros of each rational polynomial, and take their union, then we are taking a countable union of finite sets and so have a countable set. We have shown that the set of algebraic numbers is countable.

Since the union of two countable sets is countable, and the set of complex numbers is uncountable, the complement of the algebraic numbers must be uncountable. We have shown that the set of transcendental numbers is uncountable, and, in fact, that there are infinitely more transcendental numbers than algebraic ones.

13.6 Proving the Schröder–Bernstein Theorem

We have two sets X and Y and we are given injections

$$f : X \to Y \text{ and } g : Y \to X.$$

Our objective is to stitch f and g together to get a bijection from X to Y. First, note that f and g define bijections to their images, and so have well-defined inverses on their images.

$$f^{-1} : \operatorname{Im} f \to X \text{ and } g^{-1} : \operatorname{Im} g \to Y.$$

Our new bijection, h, will have to be a mixture of f and g^{-1} since these are the only two maps from parts of X to Y that we have. It is a question of how to divide up X so that all points in Y are mapped onto and there is no overlap in the images.

Since f maps no element of X to points in $Y - \operatorname{Im} f$, we will need to use g^{-1} to get those points. So we will have to define h to equal g^{-1} on $g(Y - \operatorname{Im} f)$. More generally, we will partition X according to how the points behave under the mappings.

What sort of points in X are there? We have already seen one special subset: points that map under g^{-1} to points outside the image of f. We can think about how points map under repeated application of $g \circ f$. Consider a point x, we repeatedly apply $g \circ f$ to it to get new points of X. If we take $(g \circ f)^l(x)$ for $l = 1, 2, \ldots, \infty$, we get a sequence of points. This sequence will form a loop or go on forever. We can say that points are related if you can get from one to another using $g \circ f$. More formally, we define an equivalence relation on X that two points x_1 and x_2 relate if and only if there exists $k \in \mathbb{N}$ such that

$$x_1 = (g \circ f)^k(x_2) \text{ or } x_2 = (g \circ f)^k(x_1).$$

This is easily checked to be an equivalence relation.

We now have a partition of x. Each partition set will be an infinite sequence or a loop. The infinite sequences come in three varieties. To see this, first say that a sequence is generated by x_0 if all its elements can be written in the form

$$(g \circ f)^l(x_0).$$

We can thus distinguish between sequences that are generated and so have a starting point, and those that go on forever in both directions. Note that if a sequence does not have a starting point then given any point, x, in it, there exists a point x_2 such that

$$x = (g \circ f)^m(x_2)$$

for some m. Repeating this argument, one can go back as far as one wants.

If a sequence does have a starting point, x_0, then we know that

$$x_0 \neq (g \circ f)(x)$$

for all x in X. This can happen in two ways either

$$x_0 \neq g(y)$$

for all y in Y, or, x_0 is in the image of g and we have

$$g^{-1}(x_0) \neq f(x)$$

for all $x \in X$.

We thus have a partition of X into four sets. Each point is in one of the following

(1) a loop,
(2) a doubly infinite sequence,
(3) a sequence that starts outside the image of g,
(4) a sequence that starts in the image of g.

The great thing about this partition is that applying $g \circ f$ preserves each of the four subsets. We now define h to be f in cases 1, 2, and 3, and g^{-1} in case 4. Note that all of case 4 will be in the image of g since the image of $g \circ f$ is smaller than that of g.

We still have to show that h is a bijection. First, we show that h is surjective. given $y \in Y$. Consider $g(y)$. If this in case 4, then

$$h(g(y)) = g^{-1}(g(y)) = y$$

and we are done. If not, we are in one of the three other cases and

$$g(y) = (g \circ f)(x)$$

for some x in case 1, 2 or 3. Since g is injective, $y = f(x)$ and since we are not in case 4, $y = h(x)$. We have shown that h is surjective.

It only remains to show that h is injective. Since f and g^{-1} are injective where defined, we have to show that h does not use the two functions to map different points to the same point. That is we must show that we cannot have x_4 in case 4 and x in case i for some $i < 4$, such that

$$f(x) = g^{-1}(x_4).$$

If this were to hold, then we would have

$$x_4 = (g \circ f)(x).$$

So x_4 and x would have to be in the same part of the partition by the definition of our equivalence relation, and therefore could not be in different cases, and so we are done.

13.7 Problems

Exercise 13.1 Suppose X and Y are the natural numbers. We define f to be multiplication by two from X to Y. We define g to be multiplication by three. What is the bijection h constructed by the Schröder–Bernstein theorem? What if both maps are multiplication by 2?

Exercise 13.2 Show that every real number can be represented by a decimal that goes on forever.

Exercise 13.3 A submarine starts at integer point n. It travels with integer speed k. So after turn m, its location is $n + mk$. A ship which does not know n and k drops a depth charge once a turn on an integer. Show that it is possible to design an algorithm so that the submarine is always hit eventually. What if the speeds are rational? What if they are real?

Exercise 13.4 Consider the set of real-valued functions on the reals. What is the cardinality of this set?

Exercise 13.5 Let X be an infinite set show that $X \cup \mathbb{N}$ has the same cardinality as X.

Chapter 14
Connectedness and the Jordan Curve Theorem

14.1 Definitions

An important idea in topology is that of connectedness. There are many ways to define it. Here we use a simple definition that agrees with more complex ones for open[1] subsets of $\mathbb{R}^2$. We will use it later in the book to prove that the Euler characteristic is the same for all convex polyhedra.

Definition 14.1 We shall say a subset E of $\mathbb{R}^n$ is polygonally connected if given any two points p, q, of E there exists a path consisting of a union of straight line segments contained in E that joins p to q.

So if we can draw a straight line path with only a finite number of direction changes between any two points of E then E is connected. We will call such a path a *polygonal path* and call the points of direction change *vertices*. Generally, even if a set is not connected, it can be written as a union of connected subsets. For example, let S be the unit circle in the plane, $\mathbb{R}^2$. The complement of S is not connected. Since a point inside the unit circle cannot be joined to one outside without crossing S. However,

$$\mathbb{R}^2 - S = \{||x|| < 1\} \cup \{||x|| > 1\},$$

and both of these sets are connected.

How can we prove that sets are connected other than by checking the definition? Typically, we build them up from simpler connected sets. First, we can show

Theorem 14.1 *Any convex set is connected.*

Proof If C is convex then any two points of C are joined by the straight line segment between them. This defines a polygonal path between them and we are done. □

---

[1] A subset U is open if every point in U is surrounded by points in U. More formally, if $x \in U$ there exists $\delta > 0$ such that $N_\delta(x) = \{y : |y - x| < \delta\} \subseteq U$.

© Springer International Publishing Switzerland 2015
M. Joshi, *Proof Patterns*, DOI 10.1007/978-3-319-16250-8_14

More interestingly, if we take two sets that are connected and have non-empty intersection then their union is connected.

Theorem 14.2 *If A and B are connected subsets of $\mathbb{R}^n$ and $A \cap B$ is non-empty then $A \cup B$ is connected.*

Proof Let $p, q \in A \cup B$. Let $r \in A \cap B$. Then there exists a path γ_1 from p to r and a path γ_2 from r to q. Join these two together to get a path from p to q. Any two points of $A \cup B$ are thus joinable by a polygonal path so it is connected as claimed.   □

Now suppose we have the plane with a big square subtracted. How can we show this set is connected? Suppose the square is centred at the origin and has sides of length $2R$. So let

$$E_R = \{(x, y) \mid |x| > R \text{ or } |y| > R\}.$$

We can write E as a union of four sets

$$\{x < -R\}, \ \{y < -R\}, \ \{x > R\}, \ \{y > R\}.$$

Each of these four sets is connected since they are all convex. Each one has a non-empty intersection with the preceding and succeeding sets. So add them one at a time to the union, and we see that E is connected.

For a more complex example, suppose C is convex and *bounded*. We show that $\mathbb{R}^2 - C$ is connected. We need to define bounded. We take it to mean that C lies within some square centred at the origin. So there exists R such that for all points $(x, y) \in C$, $|x| < R$ and $|y| < R$. Since we know that E_R is connected, all we need to do is to construct a straight line from any point in $\mathbb{R}^2 - C$ to E_R. This straight line can then be used along with a polygonal path in E_R to join the point to any point in E_R.

Let $p \in \mathbb{R}^2 - C$. If p is in E_R we are done. If not, consider the horizontal line through p. This gives two possible lines from p to E_R according to whether we go left or right. We claim that one of these must not intersect C. For if they both do then there is a point q on the left segment in C and a point r on the right segment also in C. But then the convexity of C would then imply that p is in C too, so this cannot happen. We therefore take the line segment from p to E_R which does not intersect C and we are done.

14.2 Components

As mentioned above, it is always possible to divide a set into connected components. The main result is

Theorem 14.3 *Suppose $E \subset \mathbb{R}^n$ then there exists a collection of subsets $\{E_\alpha\}$ whose union is E such that each E_α is connected and the subsets $\{E_\alpha\}$ are pairwise disjoint. Any connected subset of E is contained in E_α for some α.*

The term pairwise disjoint means that if $\alpha_1 \neq \alpha_2$ then

$$E_{\alpha_1} \cap E_{\alpha_2} = \varnothing.$$

Before proceeding to the proof, note that the theorem does not say anything about the cardinality of the number of components. In particular, it does not say that there is only a finite number. We will prove the result using equivalence relations.

Let $E \subset \mathbb{R}^n$. For $p, q \in E$, let $p \sim q$ if there exists a polygonal path in E from p to q. Clearly, $p \sim p$ by taking the path to be a single point. If $p \sim q$, then there is a path from p to q. Reversing the path to go from q to p yields $q \sim p$. If $p \sim q$ and $q \sim r$ then take the path from p to q and join it to the one from q to r to get a path from p to r. So $p \sim r$.

Let $\{E_\alpha\}$ be the equivalence classes defined by $\sim$. They are connected trivially by our definition. Since they are equivalence classes, they form a partition and so they are pairwise disjoint. If a subset, C, of E is connected and $p \in C$ is in E_α, then all points of C can be joined to p by a polygonal path. This means that they are in the equivalence class of p so C in contained in E_α as claimed. We have proven the theorem.

A useful alternative way to look at components is in terms of integer-valued continuous functions. This almost sounds like a contradiction since continuous functions never jump, and the only way to get from one integer to the next is by a jump. However, when working with disconnected sets the concept is not trivial.

Definition 14.2 Let $E \subset \mathbb{R}^n$. We shall say that $f : E \to \mathbb{Z}$ is polygonally continuous if it is constant on any straight line segment contained in E.

For example, suppose E has n components called $E_1, E_2, \ldots, E_n$. We let f take value j on E_j. The function f is then polygonally continuous since if any two points can be joined by a straight line segment they are in the same component and so take the same value.

In fact, a set E is connected if and only if every polygonally-continuous integer-valued function is constant. To see this, suppose p and q are such that $p \sim q$ and g is continuous. We have a polygonal path from p to q and along each segment in the path g is constant so it must be constant on the entire path. The value at p and q is therefore the same. Hence, g is constant on each component of E.

If E is connected then there is only one component and the function is constant. If E is not connected then there are at least 2 components. Let g take the value 0 on the first component and 1 elsewhere. The function g is continuous and so there is a non-constant integer-valued polygonally continuous function.

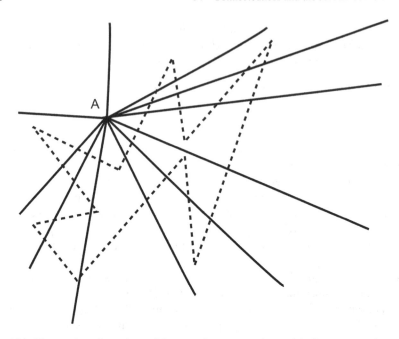

Fig. 14.1 The number of crossings of the *curve* by a ray starting at *A* is always even

14.3 The Jordan Closed-Curve Theorem

Take a closed loop in the plane with no self intersections. Thus we have a path, P, that meanders around and returns to its starting point without touching itself in between. Such a path is said to be *closed*. Consider the set $\mathbb{R}^2 - P$, how many components does it have? The Jordan closed-curve theorem says that the answer is always 2. Whilst the result is obvious, that does not mean it is easy to prove. Here we look at how to prove it for polygonal paths. The proof for general continuous paths is hard and far beyond our scope.

There are really two parts to the proof. First, we show that there are at least two components and then we show that there at most two components.

Lemma 14.1 *Let P be a closed polygonal path without self intersections, then $\mathbb{R}^2 - P$ has at least two components.*

Proof We construct an integer-valued polygonally continuous function that takes two values. Given a point p in $\mathbb{R}^2 - P$, take a ray l from p. A ray is an infinite half straight-line. Think of a laser beam emanating from p in some direction and going off to infinity. The ray l may intersect P. Its intersection will be a union of points and straight line segments. Let x be one of these points or line segments. We call x a crossing of P if the part of P before and after x lie on different sides of l. Call the total number of crossings $c(p, l)$.

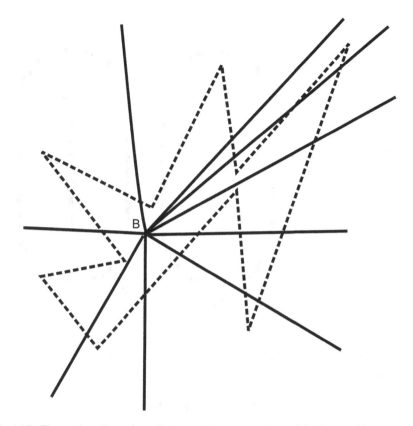

Fig. 14.2 The number of crossings of the *curve* by a ray starting at *B* is always odd

There is no reason to think that $c(p, l)$ will the same for all rays starting at l. In fact, it is easy to construct examples where it is not. A ray in some direction may miss *P* altogether whilst in the opposite direction it may hit *P*. However, if we slowly rotate l, we see that it only changes when l hits a vertex of *P* and it then stays the same or goes up or down by 2. So $c(p, l)$ varies with l, but whether it is odd or even does not. We therefore define $c(p)$ to be the remainder of $c(p, l)$ on division by 2. See Figs. 14.1 and 14.2.

Now if two points p, q, are connected by a straight line in $\mathbb{R}^2 - P$ then we can extend that line into a ray from p that goes through q. Since there are no points in *P* on the part of the ray from p to q, we see that $c(p)$ and $c(q)$ are the same. So c is indeed an integer-valued polygonally continuous function.

We still need to show that c is not constant. Take a point r in *P* and a short line segment through r that crosses *P* and does not meet *P* anywhere else. Let p and q

Fig. 14.3 The sets N_E and Q_E

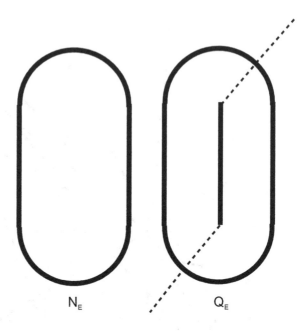

N_E Q_E

be the two endpoints of it. Let l be the ray from p through q and let l_q the part of that ray after q. We clearly have

$$c(p, l) = c(q, l_q) + 1.$$

So

$$c(p) \neq c(q).$$

$\square$

Lemma 14.2 *Let P be a closed polygonal path without self intersections, then $\mathbb{R}^2 - P$ has at most two components.*

Proof Before proceeding to the main part of the proof, we note some facts about vertices and edges. First, if a vertex is not part of an edge then since the edge is a closed set, we can find $\epsilon > 0$ such that all points in the edge are more than ϵ away from the vertex.

Second, given two disjoint edges the point of closest approach will be a vertex of one of the two edges. To see this, observe that either the two edges are parallel or they are not. If they are, then it is possible to slide two points along them without changing distance until one is a vertex. So the vertex is as close as anything to the other edge. If they are not parallel then one can slide along in the direction in which they get closer together until one is a vertex. So the pair of closest points must have one as a vertex.

Now pick δ so that every vertex is at least 2δ away from every edge of which it is not a member. We can do this since there are only a finite number of vertices. Then every edge is at least 2δ away from every edge it does not intersect by the arguments above.

Given an edge E, let N_E be the points less than δ away from E, and let Q_E be the points in N_E that are not in the polygonal path. See Fig. 14.3. The set N_E will be a rectangle with a semi-circle added on top and bottom. The radius of both semi-circles will be δ and the rectangle will have short side 2δ and long side equal to the length of E. Subtracting P takes a middle vertical line out of the centre of the rectangle. It also takes a radial straight line (like a spoke) out of each of the two semi-circles. These clearly divide Q_E into two connected sets. These two sets will have different parity functions according to the component function, c, we defined when proving that there are at least 2 components. The sets N_E will only intersect when the edges they surround intersect due to the fashion in which we picked δ.

If we now consider the set of all points, P_δ, within δ of P then it will be a union of the sets Q_E. Taking a particular edge E, each of the two components of Q_E will intersect with the component of the same parity for each neighbouring edge, and so its union with it is connected. Going round the polygon, we see that P_δ has two connected components.

Now consider an arbitrary point $p \in \mathbb{R}^2 - P$. We will show that it can be straight-line connected in $\mathbb{R}^2 - P$ to one of these two components. Take a ray starting at p. Rotate it until it hits P. It must hit P_δ before P, since all points less than δ from P are in P_δ. So we have a straight line from p to one of the two components of P_δ and we are done. $\square$

14.4 Problems

Exercise 14.1 Show that a subset of $\mathbb{R}^n$ has k components if and only there exists an integer-valued continuous function that takes k different values.

Exercise 14.2 Show that if a point, p, can be joined to a connected set, U, by a polygonal path then p and all points of U are in the same component.

Exercise 14.3 Will a finite intersection of connected sets be connected?

Exercise 14.4 Suppose we have a collection of connected sets U_j such that for all $j > 1$, there is some $k < j$, such that

$$U_j \cap U_k \neq \varnothing,$$

does this guarantee that $\bigcup_j U_j$ is connected?

Exercise 14.5 Suppose we take the real line and subtract k distinct points, how many components will there be? Prove it.

Chapter 15
The Euler Characteristic and the Classification of Regular Polyhedra

15.1 Introduction

It has long been known that there are only five regular polyhedra. These are the tetrahedron, the cube, the octahedron, the dodecahedron and the icosahedron. How can we define a regular polyhedron and how can we prove this list is complete? The usual definition is that every face must have the same number of edges, and the same number of faces must meet at every vertex.

The key to most proofs of this classification is the *Euler characteristic*. This tells us how to compute the number of edges, E, given the number of faces, F, and the number of vertices, V.

For all the regular polyhedra, we get

$$V - E + F = 2.$$

See Table 15.1, the quantity $V - E + F$ is called the *Euler characteristic*. It takes the value 2 for many more polyhedra than just the regular ones. In this chapter, we will explore when it takes the value 2, how to prove it takes the value 2 and then how to use this fact to classify the regular polyhedra. We will use various proof patterns en route.

When mathematicians discuss polyhedra they generally mean the surface of a solid polyhedron. We will use the term both for the surface and the solid object in this chapter.

15.2 The Euler Characteristic and Surgery

The first thing to note about the Euler characteristic is that it is very robust. We can do all sorts of things to a polyhedron without changing its characteristic. As a first operation, consider drawing a line from a vertex to another non-adjacent vertex on

© Springer International Publishing Switzerland 2015
M. Joshi, *Proof Patterns*, DOI 10.1007/978-3-319-16250-8_15

Table 15.1 The Euler characteristic for regular polyhedra

Shape	V	E	F	V − E + F
Tetrahedron	4	6	4	2
Cube	8	12	6	2
Octahedron	6	12	8	2
Dodecahedron	20	30	12	2
Icosahedron	12	30	20	2

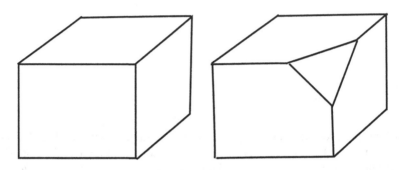

Fig. 15.1 A cube before and after chopping off a corner

the same face. Call this line an extra edge. We have divided the face into 2 pieces so we also have an extra face. The number of vertices has not changed. So $V - E + F$ has not changed either. In conclusion, dividing a face in two by drawing a line between two non-adjacent vertices does not change the Euler characteristic.

Now suppose we chop off a corner. We have a vertex where k edges meet and we replace it by a small face. See Fig. 15.1. The new face will have k sides because there are k edges coming in. It will also have k vertices since a face must have the same number of sides and vertices. For the polyhedron, we have increased the faces by one, the edges by k and the number of vertices by $k - 1$. So $V - E + F$ changes by

$$k - 1 - k + 1 = 0.$$

Chopping off a corner does not change the Euler characteristic.

We can also deform polyhedra: if we move a vertex and stretch the edges emanating from it correspondingly, then the number of vertices, edges and faces does not change and so neither does the Euler characteristic.

We can also glue polyhedra together. Suppose we have two polyhedra P_1 and P_2 with Euler characteristics χ_1 and χ_2. Let the number of vertices, edges and faces in P_j be V_j, E_j, F_j. Suppose we can find a face in P_1 which is identical to one in P_2. We now glue those two faces together to get a new polyhedron. What is the new Euler characteristic? Suppose the joined-together faces have k edges and k vertices. The number of vertices in the new polyhedron will be

$$V_1 + V_2 - k.$$

The number of edges will be

$$E_1 + E_2 - k,$$

and the number of faces will be

$$F_1 + F_2 - 2,$$

since the two glued-together faces have disappeared. The Euler characteristic of P is therefore

$$\chi = \chi_1 + \chi_2 - 2.$$

So gluing on a shape with Euler characteristic 2 does not change the Euler characteristic.

With all these operations, we can now construct a lot of polyhedra with Euler characteristic 2. Equally, we have shown that if a polyhedron does not have Euler characteristic 2 then it cannot be constructed from regular polyhedra using these operations using *difference of invariants*.

Two obvious questions arise:

- How can we prove that a large class of polyhedra have Euler characteristic 2?
- Are there polyhedra that do not have Euler characteristic 2?

We will answer both of these in this chapter.

15.3 Transforming the Problem

We will prove that convex polyhedra have Euler characteristic 2. Recall that a subset, A, of $\mathbb{R}^n$ is convex if given any points in it, the line between them also lies in A. We first change the problem to a statement about networks in the plane.

It is easier to prove statements when every face is a triangle. We saw in Sect. 1.5 that every polygon can be triangulated by drawing straight lines between vertices. We saw above that this operation does not change the Euler characteristic. It is therefore enough to prove the result in the case that all faces are triangles. This is an example of specific generality.

We next transform into a statement about lines inscribed on a sphere. Take a large sphere which contains the polyhedron. Take a point, p, inside the polyhedron. We define a map from the surface of the polyhedron to the sphere. Let q be a point in the surface. Let $\theta_p(q)$ be the point on the sphere where the straight line starting at p which passes through q hits the sphere.

The straight line for each point q will be different because the polyhedron is convex and so each point q will get mapped to a different point of the sphere. Equally if we take a point, r, on the sphere, then the line through it and p must intersect the surface of the polyhedron somewhere and this point will be mapped to r, so we have a bijection. We thus have mapped all the edges and vertices onto the sphere, and they

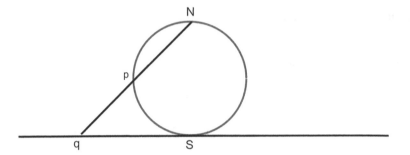

Fig. 15.2 Stereographic projection for a *circle* to a *line*

then now surround areas of the sphere which we can regard as faces. The network on the sphere and the polyhedron will have the same Euler characteristic. We therefore now only need consider the case of sphere with a network of lines and vertices drawn on it. Note the crucial property that two lines can only intersect in a vertex. If we started with a polyhedron with triangular faces then each face on the sphere will also have 3 sides and 3 vertices.

It is easier to work with networks in the plane than on a sphere so we can transform the problem further. We rotate and translate the sphere so that the south pole is at the origin in three-dimensional space, and so that the north pole is in the interior of some face. Having done so, we can now use *stereographic projection* to map the sphere to the plane. We define a map, ϕ, from the sphere minus the north pole to the plane, for a point p in the sphere, we take the straight line from the north pole to p and extend it until it hits the plane, that is the set with $x_3 = 0$. We map p to that point, q. See Fig. 15.2.

Every face other than the one containing the north pole maps to a triangular-shaped region in the plane. The sides of these may be curved, however. The north pole containing region maps to all the plane outside a large triangle which contains all the other triangles. If we do not regard the exterior region as a face, then our problem has transformed to proving that for a network of lines and vertices surrounding triangles, the value of $V - E + F$ must be 1.

15.4 The Result for Networks in the Plane

In this section, we prove that if we have a network of triangles in the plane then we have $V - E + F = 1$. We assume that the triangles can only intersect along common edges or a single vertex. We regard the network as the union of vertices, edges and faces contained in it. We also assume that the network is polygonally connected. Note that a path between two points in different triangles can pass first to the edge of one triangle, then along edges and vertices, and then to the final point. So it is the only first and last part of the path that involve triangle interiors. The result cannot be true without this hypothesis of connectedness: n disjoint triangles will have Euler characteristic n. Our final assumption is that the area outside the network is also connected. This will certainly be the case for the network resulting from polyhedra according to our projection.

We proceed by induction. Our inductive hypothesis is that ALL such networks containing n triangles have Euler characteristic 1. When $n = 1$, we only have 1 triangle and the result is clear.

Given a general such network with $n + 1$ triangles, we need to reduce to the n case and then apply the inductive hypothesis. We will remove a triangle on the edge of the network. There are a number of ways a triangle can be on the edge. It will not share 3 edges since it is on the edge, and it must share at least one vertex since the network is connected. Note also that if an edge is shared so is its endpoints. So the intersection set could be

(1) zero edges and one vertex;
(2) zero edges and two vertices;
(3) zero edges and three vertices;
(4) one edge and two vertices;
(5) one edge and three vertices;
(6) two edges and three vertices.

See Fig. 15.3 for the cases 1, 4 and 6. In case 1, we delete the triangle. We remove 3 edges, 2 vertices and 1 face. The Euler characteristic does not change. The network's complement is also clearly still connected. For the network's connectedness, see that if the path between two points in the remaining network had passed through this triangle then it must have crossed the not-deleted vertex twice: on the way into the triangle and the way out. So we simply remove the part of the path in between and we still have a path between the two points.

In case 4, we delete the triangle by removing the vertex and two edges that are not in the network. We have removed 2 edges, 1 vertex and 1 face so the Euler

Fig. 15.3 The three cases of triangles that can be easily removed

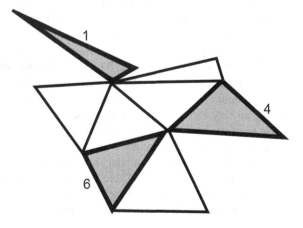

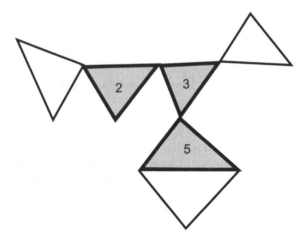

Fig. 15.4 The three cases of triangles that cannot be easily removed

characteristic is again invariant. We clearly have not changed the connectedness of the network's complement. For the connectedness of the network, we replace any part of a path inside the triangle with a path along the edge, and so it is still connected.

In case 6, the sole edge of the triangle that is not in the intersection with the rest of the network is removed. Its deletion removes the face of the triangle and the edge itself so we decrease F and E by one and the Euler characteristic is unchanged. The edge is on the boundary of the network so we are adding the interior of the triangle and the edge to the complement. The network's complement therefore remains connected. We move any path through the triangle to run along the two edges left and the network is still connected.

We are left with cases 2, 3 and 5. See Fig. 15.4. These cases are different in that a path through the triangle cannot be easily deformed to not pass through it—we can expect that deleting such a triangle does indeed disconnect the network. Our solution is therefore to find a triangle that is in one of the other cases. So pick a triangle on the boundary, if it is in another case, we are done. If it is in case 2, 3 or 5, it must be next to another triangle on the boundary. Look at that triangle. If it is not in case 2, 3 or 5, we delete it. If it is, observe that there must be a third triangle on the boundary next to it. Again if this is not in case 2, 3 or 5, delete otherwise continue. We keep going until we find a deletable triangle. We need to show that the process stops. Since there are only a finite number of triangles, it can only stop if we return to a case 2, 3 or 5 triangle already considered or we find one not in case 2, 3 or 5.

We need to show that the first does not occur. Suppose it does. We then have a loop of triangles on the boundary. These triangles are all on the boundary so they will always have an edge on the boundary and we can trace all these edges to get a closed polygonal loop. See Fig. 15.5. There will be points on both sides of this loop which are not in the network since all the triangles are in case 2, 3 or 5. We now have to invoke a big theorem: the *Jordan closed-curve theorem* from Chap. 14. It tells us that

Fig. 15.5 A bad
configuration in which the
triangles form a loop

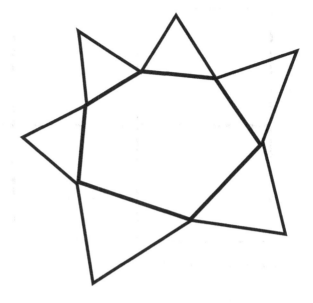

the two sides of the loop are not connected to each other. The loop of case 2, 3 and
5 triangles has made the complement of the network disconnected. This contradicts
our inductive hypothesis so such loops do not occur and we are done.

Note that we have used induction here and we have worked down rather than up.
An alternate approach might have been to start with one triangle and then stick more
on. The main trickiness would then lie in showing that you got every possible network
eventually. We also made the proof easier by assuming the inductive hypotheses for
all possible connected networks with connected complement and one less triangle
rather than trying to just assume it for the one network that was relevant.

The appearance of the Jordan closed-curve theorem is important in that it is at
this point that we make use of the fact that the polyhedron is topologically a sphere.
If our proof did not use it or something similar, we would be suspicious of it.

15.5 Counterexamples

An obvious question now arises, do all polyhedra have Euler characteristic 2?
That partially depends on the definition of a polyhedron! We construct some three-
dimensional shapes where it takes a different value. Suppose we take 5 cubes and
glue them together in a U pattern. See Fig. 15.6. So we take one, glue a second one
to its bottom, glue 2 more to its side and then glue a further one to the last one's top.
We know that this U shape has Euler characteristic 2 since it was made by gluing
cubes together, and we showed above that this does not change the value. However,

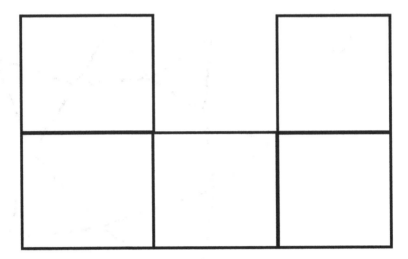

Fig. 15.6 Five cubes glued together in a "U" shape viewed from above

if we take two of these U shapes and glue them together on the faces at the tops of the U shapes then we are gluing two faces at once and we get a ring-shape.

What is the Euler characteristic? We have eliminated 2 faces, 4 edges and 4 vertices for each of the gluings, so we get

$$\chi = \chi_1 + \chi_2 + 2 \times (-4 - (-4) - 2) = 2 + 2 + 2 \times (-2) = 0.$$

The ring, or doughnut-shape, has Euler characteristic zero. The surface of this shape is generally called a torus. We have shown that Euler characteristics do vary. Note that there is a qualitative difference in this shape from the convex polyhedra: if we take a loop on a torus that goes round the ring, it is not possible to shrink it to a point without breaking it or leaving the torus. In a sphere, we can always just push the entire loop to the north pole so it is quite different.

We can repeatedly glue on more and more U shapes, joining both the ends each time. Each time we reduce the Euler characteristic by 2 and get an essentially different shape. The number of holes is called the *genus g* and we have

$$\chi = 2 - 2g.$$

For closed surfaces in three-dimensional Euclidean space, it can be shown that given an appropriate notion of deformation, two surfaces can be deformed into each other if and only if they have the same g or χ.

15.6 Classifying Regular Polyhedra

We have listed five regular polyhedra and it has long been known that this list is complete but how we can prove it? A polyhedron has V vertices, E edges, F faces, and since it is regular, suppose every face is a polygon with p sides and q faces meet at each vertex. What relations can we deduce? We must have

$$V - E + F = 2,$$
$$pF = 2E,$$
$$qV = 2E.$$

The middle equation is because every face has p edges and each edge is in two faces. The last is because q edges meet at each vertex and each edge has two vertices. We have three equations and all the solutions must be integers. We can eliminate V and F from the first equation, to get

$$\frac{2E}{q} - E + \frac{2E}{p} = 2,$$

or

$$\frac{1}{p} + \frac{1}{q} = \frac{1}{2} + \frac{1}{E}.$$

The number of edges must be positive so we have

$$\frac{1}{p} + \frac{1}{q} > \frac{1}{2}.$$

Now p and q are positive integers. A polygon must have 3 sides and in a polyhedron at least 3 sides must meet at a vertex so they are both at least 3. Since

$$1/p \leq 1/3$$

we have

$$1/q > 1/6.$$

This means that q is 3, 4 or 5. The same argument shows that p is 3, 4 or 5. We now have nine pairs, checking them, we see that the only possibilities for the pair (p, q) are the five cases

$$(3, 3), \ (3, 4), \ (3, 5), \ (4, 3), \ (5, 3).$$

These yield the five polyhedra we listed initially and we are done.

15.7 Problems

Exercise 15.1 Suppose we make two W shapes out of cubes and join the three top bits together what happens?

Exercise 15.2 If we attempt to apply our proof that the Euler characteristic is 2 to a ring shape, where does it fail?

Exercise 15.3 Suppose we take a square in the plane and cut it into small triangles, what is the Euler characteristic? Suppose we stick opposite sides together, what is the new Euler characteristic? What if we twist one pair of sides before gluing?

Exercise 15.4 Suppose we take a regular polyhedron and form a new one by making each vertex the centre of one of the old faces and joining vertices from neighbouring faces with edges. What happens for each of the 5 regular polyhedra?

Chapter 16
Discharging

16.1 Introduction

In this chapter, we use the notion of discharging to prove more results about polyhedra and maps. First, we give an alternative more recent proof that the Euler characteristic is always 2 for a convex polyhedron, and then we look at how to combine discharging and double counting to prove that only certain face combinations are possible for any polyhedron.

In order to discharge a polyhedron, one first has to charge it. The notion is similar to that of electric charge. We put one unit of charge on every face and vertex, and minus one units on every edge. The Euler characteristic is then the total charge on the polyhedron.

16.2 The Euler Characteristic via Discharging

We present a proof due to William Thurston that the Euler characteristic of certain polyhedra is 2. We suppose that the network defined by the edges can be inscribed on a sphere without changing the number of faces, edges and vertices and their intersections. We saw that this was possible for convex polyhedra in the last chapter. We also saw previously that we can make all faces triangular without affecting the Euler characteristic so assume also that this has been done.

First, we rotate the sphere and deform the network so that one vertex is at the north pole and another is at the south pole. We also deform the network so that no edge is horizontal: we make sure that no two vertices have the same latitude. Now charge the network as discussed above. We push charge eastwards. We make the charge on each edge go on to the face to its east. Since no edge is horizontal the eastwards face is well defined. If a vertex is between two edges both going on the same face its charge goes to that face as well. For other than the two poles, since every vertex

© Springer International Publishing Switzerland 2015
M. Joshi, *Proof Patterns*, DOI 10.1007/978-3-319-16250-8_16

is surrounded by triangles which have it as a vertex, this will be true for exactly one of those triangles.

Each face therefore gains the charge from either one edge and so has net charge zero, or gains the charge from two edges and one vertex and again has zero net charge. At the end, the only charge remaining is that on the north and south poles and we have 2. The charge-moving procedure has not changed the net charge so the Euler characteristic must have been 2 initially and we are done.

16.3 Maps and Double Counting

We saw in Chap. 12 that the Four-Colour theorem in general is implied by the special case that three countries meet at every vertex. Such a map is sometimes said to be *cubic*. Counting the area outside the map as a single extra country, we can use stereographic projection to change a map into a polyhedron. What can we say about cubic polyhedra? We can count up the edges, faces and vertices in terms of how many sides each face has. Thus let N_k be the number of faces with k sides. We then have

$$F = \sum_k N_k.$$

For edges, we get

$$2E = \sum_k k N_k,$$

since each edge is two sides, and if a face has k sides that is the same as saying it has k edges. For vertices, since the map is cubic, each vertex is in 3 faces. So

$$3V = \sum_k k N_k.$$

We know

$$V - E + F = 2.$$

We conclude

$$\sum_k (1 - k/2 + k/3) N_k = 2.$$

equivalently,

$$\sum_k (6 - k) N_k = 12.$$

This immediately says that for some $k < 6$, $N_k > 0$: there must be at least one side with less than 6 sides.

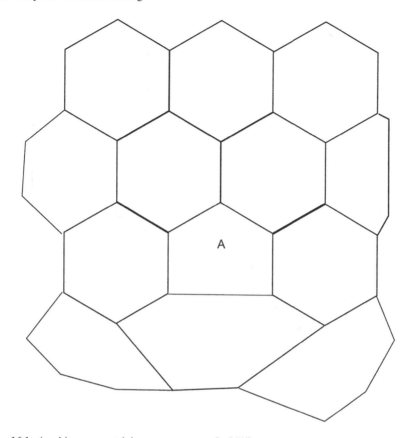

Fig. 16.1 A cubic map containing a pentagon marked "A"

In fact, the result does quite a bit better. Each face has at least 3 sides so we must have 4 faces with less than 6 sides. If each face has at least 4 sides, then we must have at least 6 faces with 4 or 5 sides. If there are no faces with less than 5 sides then we must have 12 faces with 5 sides. A dodecahedron has 12 faces with 5 sides and 3 faces meet at each vertex so this is as good as you can get.

We can now easily prove the Six-Colour theorem.

Theorem 16.1 *Any map of the plane can be coloured with 6 colours so that no two territories with a common edge have the same colour.*

Proof It is enough to consider the cubic case. We proceed by complete induction. If there are less than 7 countries, the result is clear. Suppose every map with n or fewer countries can be done. If our map has $n + 1$ countries, find a country with 5 or fewer sides. See Fig. 16.1. Merge this country with one of its neighbours. The new map has n countries so it can be six-coloured. Now restore the country that was merged. See Fig. 16.2. It only has 5 or fewer neighbours, so assign it a colour that none of its neighbours have and we have a colouring. Unfortunately, this argument

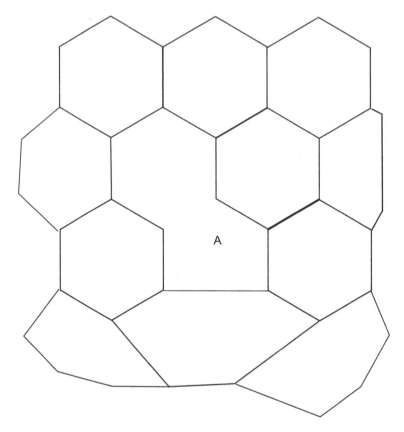

Fig. 16.2 A cubic map in which the pentagon marked "A" has been merged with an adjacent hexagon

has a flaw. We have implicitly assumed that the two countries merged do not have more than one side in common. If the two sides in common are adjacent then they are not really separate sides and we do not have a problem. If they are not adjacent then post merging the new country is a ring shape and so we have a map interior to it and one exterior. By considering the ring and its exterior as a single country, we can colour the interior and the ring with 6 colours. Rearrange the colours so that the ring is yellow. Now do the same for the exterior and the ring. Since both colourings make the ring yellow, we can merge the two and have a colouring of the map including the ring. We can reinsert the pentagon, as before and give it the one colour that does not neighbour it.

The result follows by complete induction.   □

Of course, we would really like to prove the Four-Colour theorem not the Six-Colour one. This is much harder but the basic approach is the same, show that certain configurations of territories are inevitable and that they can be handled.

16.4 Inevitable Configurations

In the last section, we saw that for a cubic map, there will always be a face with less than 6 sides. We can actually do better. We can prove that there must always be a face with less than 5 sides or a face with 5 sides adjacent to a face with 5 or 6 sides. The proof uses discharging.

Suppose every face has at least 5 sides. To each face with k sides, assign a charge of $6 - k$. We know

$$\sum_k (6 - k) N_k = 12.$$

The only faces with positive charge have 5 sides. Suppose each one only has neighbours with seven or more faces. Take the positive charge and share it equally among the neighbours. We show that the neighbours will continue to have negative charge. If a face has $2k$ edges then it has charge $6 - 2k$ before the discharging. It can have at most k pentagon neighbours if no two are adjacent, so it can gain at most $k/5$ charge. So its final charge is at most

$$6 - 2k + k/5$$

which is less than zero for $k > 3$. If it has $2k + 1$ faces it can again have at most k pentagon neighbours so its final charge is at most

$$6 - 2k - 1 + k/5 = 5 - 1\frac{4}{5}k,$$

which is again less than zero for $k > 3$. We thus have that all faces have zero or negative charge. This is impossible if the charges add up to 12. We conclude that some pentagonal face has a neighbour with five or six sides.

16.5 Problems

Exercise 16.1 What happens if we attempt to apply Thurston's discharging proof to a torus? (i.e. the surface of a doughnut.)

Exercise 16.2 Is it possible to have a polyhedron whose faces are all pentagons or hexagons? How many pentagons must such a polyhedron have? Suppose we merge two adjacent faces what happens?

Chapter 17
The Matching Problem

17.1 Introduction

The Principal of Silly College has decided that his students are spending too much time dating and not enough time studying. He decides that the easiest way to solve the problem is to pair off the students as couples in such a way that everyone is happy enough to make the pairings stable. He therefore summons the college mathematicians, Jack and Jill, and asks them to devise an algorithm for pairing the students.

17.2 Formulating the Problem

First, the problem has to be given a mathematical formulation. Jack and Jill arrive at the following formulation.

- There are N male and N female students.
- Each student is required to provide a list of the students of the opposite sex in order of desirability with most desirable first.
- A pairing is *stable* if there does not exist a male and female who prefer each other to their assigned partners.

Note that this definition of stability does not require anyone to have their ideal mate, nor does it have a concept of optimality; there may be other pairings that make more students happy in some sense.

Once a definition has been formulated, three obvious questions are generally asked by mathematicians:

(1) does a stable pairing always exist?
(2) are stable pairings unique?
(3) how can we find a stable pairing?

We now see how to answer all of these using an algorithmic construction.

© Springer International Publishing Switzerland 2015
M. Joshi, *Proof Patterns*, DOI 10.1007/978-3-319-16250-8_17

17.3 The Algorithm

Jack comes back the next day with an algorithm.

(1) Every male student who is not paired, proposes to the highest ranked female on his list who has not yet said no to him.
(2) Each female who has multiple offers takes the one she ranks highest. She does this even if she has already accepted an offer from someone who is lower on her list. If her new offers are worse than her previously accepted offer she ignores them.
(3) This process is repeated until everyone is paired.

For this algorithm to be useful, we have to show that it always terminates and that the resulting pairing is indeed stable. This algorithm is called the *Gale–Shapley algorithm.*

To see that the algorithm always terminates, observe that once a female is engaged she is always engaged (but possibly to someone else,) and that once she has had one proposal this is always the case. Observe also that every female will always get at least one proposal since each male will work down his list until he is paired, and as long as there is a surplus female there is also a surplus male.

We now show that the resulting pairing is stable. Consider a male A and a female B who are not paired to each other. We must show either that A prefers his current partner to B, or that B prefers hers to A. If A does not prefer his current partner to B, then he will have asked B before his current partner. She must have turned him down for someone else she preferred or he would be paired with her. Given that she only changes partners when she gets a better offer, she therefore prefers her current partner to A. So the pairing is indeed stable.

17.4 Uniqueness

The question of uniqueness remains. Immediately, after Jack presents his solution, Jill derides it as sexist since men get to do all the proposing. She therefore switches the role of males and females. Since the problem formulation does not change when we switch them, her algorithm is equally valid. We now have two different algorithms that produce stable matchings.

The existence of two different algorithms alone does not imply non-uniqueness. They could always end up at the same solution. Indeed, this is often the case in computer science where the objective of finding new algorithms is to reach the same solution faster rather than to find a different one. However, if we present input data that causes the two algorithms to lead to different outcomes then we do indeed have non-uniqueness. Call the male students A, B and C and the female students X, Y and Z.

Suppose the lists are as follows:

- $A : XYZ$.
- $B : YZX$.
- $C : ZXY$.

Then it does not matter what the female lists are for Jack's algorithm. Each male gets his first choice and we are done. A gets X, B gets Y and C gets Z. Now suppose the female lists are

- $X : CBA$.
- $Y : BAC$.
- $Z : ACB$.

and we use Jill's algorithm, then X gets C, Y gets B and Z gets A. Since both pairings came from algorithms guaranteed to produced stable results, we now have two distinct stable pairings. We have demonstrated non-uniqueness.

We have shown the existence and non-uniqueness of stable pairings by algorithmic construction. In designing these algorithms, we have not put in any form of optimality criterion. All we have looked for is stability. One pairing may be better than another in that it leads to greater happiness but we have not built that consideration into our algorithms.

17.5 Further Reading

For more discussion of the matching problem and related questions, see Gura and Maschler's "Insights into game theory."

17.6 Problems

Exercise 17.1 The room-mate problem is to pair new students of the same sex into rooms. A pairing is stable if no two students prefer each other to their room-mate. Does the algorithm presented here apply?

Exercise 17.2 We can rate a matching of all students by summing all the ranks of each member of a pair for his/her partner. Show that there is a stable pairing which minimizes this rating amongst stable pairings. Will it be unique? Design an algorithm to find it.

Chapter 18
Games

18.1 Introduction

In this chapter, we look briefly at game theory. In particular, we look at how one might use mathematics to answer questions such as is it ever advantageous to play second in a game? In answering that question, we see an example of a non-trivial application of proof by contradiction.

What is game theory? It is the study of how to find optimal strategies for playing games. It is often applied to situations well outside board games such as negotiations and evolution. Here we will stick to "classic games" in which there are two players, they play alternately, there is no luck and at all times both players know the complete state of the game. Examples of such games are chess, draughts, go, othello, abalone, Nim, and noughts and crosses.

We look at the following questions

- must there exist an optimal strategy?
- must games always end?
- are there games where we can be sure that going second is bad?

18.2 Defining a Game

There are many ways to define a classical game. Here we define it using a state-space and relations that specify which moves are legal. Thus we can define a game-state to be a finite sequence of integers or real numbers. Note that our objective here is to prove theorems, not to develop efficient computer algorithms so our data structures are ones that would be unlikely to be used by a software engineer. For traditional games, integers will suffice. For example, the state of chess board can be described as a sequence of 64 integers each one denoting what piece is in that square and a zero denoting an empty square. A "go board" is 19^2-squared lattice points each one containing 0 for nothing, 1 for a black piece and 2 for a white piece.

© Springer International Publishing Switzerland 2015
M. Joshi, *Proof Patterns*, DOI 10.1007/978-3-319-16250-8_18

In noughts and crosses, also known as tic-tac-toe, the players take turns to write down their symbol, a cross or a nought, into empty squares until all squares are filled or the victory condition of 3 in a row is achieved. More generally, we could consider an n-dimensional board of k squares in each direction with a victory condition of achieving l in a row. We will call this (n, k, l) noughts and crosses.

Both chess and go have a rules subtlety designed to prevent infinite loops. In chess if the same position occurs 3 times in the same game, a draw can be claimed. In go, it is illegal to return the board to a state that has been previously achieved. Thus the state of the game is not just the state of the board but also all previously achieved board-states. We can easily encapsulate this mathematically by defining the game state to be the vector of all board states so far achieved in the game. We will also generally want an extra state variable to specify termination and who won, for example white wins, black wins, draw, game not over yet.

The game Nim has a very simple state-space. In this game, two players take turns to remove 1–3 matches from a pile of them. The person who takes the last match wins. The state space is simply a natural number representing the number of matches left in the pile.

We can define the game rules to be a relation, R, between game states. Thus the legal moves from a position x are simply the positions y such that $x R y$.

Having defined games mathematically, we can now start trying to prove theorems.

18.3 Termination

We show that chess, go and (n, k, l) noughts and crosses must terminate. Noughts and crosses is the easiest. The board has k^n squares and every turn one is filled in. After k^n turns the board is full, and the game is over so noughts and crosses certainly ends.

Now chess, we can prove termination by using the three-repetition rule. Each player has 6 different sorts of pieces, and a square must contain one of these or be empty so there are at most 64^{13} legal positions. Eventually, one of these must be attained 3 times so the game must end. Our upper bound on the numbers of moves is rather crude, and chess generally has other ways to ensure termination in a reasonable amount of time. However, our approach is sufficient to prove that the game ends.

In go, similarly, we have 3^{261} possible board states. So eventually the no-repetition rule will ensure that neither player has a legal move and they will both have to pass and the game is over.

In Nim, we can use a simpler argument, if there are n matches and each player always takes at least one, the game will be over in at most n turns.

18.4 Optimal Strategy

We can ask the question of who will win if both players play optimally. For example, in Nim there is a well-known strategy. If there are $4k + l$ matches left with k an integer and l between 1 and 3 inclusive, take l matches. If $l = 0$ and it is your turn, then you have lost, so take 1 and hope your opponent does not play optimally. This strategy works because after your turn there is always a multiple of 4 matches left. The number of matches goes down by 4 every two turns and so eventually hits zero with you having taken the last match.

The game Nim is said to be "first player wins" if you start with a non-multiple of four matches, and "second player wins" otherwise. We would like to classify other games as "first player wins", "second player wins" or "draw". Before doing so we really need to show that every classical game does indeed have an optimal strategy.

We can proceed by backwards induction. There are a finite number of finished game board states. In each of these, the victor or draw is known. We label them accordingly. We next consider positions where all moves put you into an already labeled position, or the player moving can achieve a position labeled with their own victory. We label all these positions to be the best outcome they can achieve so if they can get a victory then the player moving wins; if not and they can achieve a draw then "draw"; otherwise label with other player wins.

Once this is done. We repeat until all positions have been labeled. Note that each time we label a position, more positions will have all possible moves labeled. To see who wins we look at the initial position and we are done.

We really ought to show that the process will always terminate in the sense that we will not reach a point where no positions can be labeled. However, if we know that the game must terminate after N moves, then at the kth stage we must have labeled all games that have taken $N - k$ moves. So we will finish labeling when $k = N$ at the latest.

Our optimal strategy is now just to play the state that has the same label as the current board state. If there are multiple such states then choose any one. This would then indicate that there is more than one optimal strategy.

Our argument has shown that an optimal strategy exists and shown how to construct it. Unfortunately, for reasonably complex games it is not feasible to program in that the amount of computational power required would be immense. Indeed, it is only recently that computer programs that play go reasonably have been implemented, and they are still a long way from defeating human professionals.

18.5 Second Player Never Wins

In some games, one can prove that optimal play will not make the second player win. The essential feature of such games is that there is never a disadvantage to playing. This is not the case for chess where a special word "zugzwang" is used to denote a position where the optimal move would be no move if that were legal. It is also clearly not the case in Nim where second player does win if there are 4 sticks left.

In noughts and crosses, however, it is always preferable to play. The reason being that having an extra piece on the board in no way restricts you or makes your position worse. We can use proof by contradiction to prove that there is not a strategy to make the second player always win. Suppose such a strategy existed. The first player would make a random first move, and then would be in the position of the second player with an extra piece on the board. From then on, he would follow the second player's optimal strategy. If the strategy was to play at a point already occupied by his extra piece then we just plays an extra piece anywhere on the board. Since this strategy makes the second player win, it must also make the first player win since the extra piece can cause no harm. We have a contradiction: if second player always wins then first player always wins. We conclude that with optimal play either first player wins or there is a draw.

18.6 Problems

Exercise 18.1 Suppose Nim is played with 2 stacks and each player can only take from one of them each turn. Analyze this game.

Exercise 18.2 Suppose we play a version of Nim on a clock face. We start with the hour hand at 12 the first player to get it to 6 wins. Each turn a player can move the hour hand 1–3 integer hours. What happens? What if we add in the rule that the clock cannot show the same time twice? What if we delete any time that has been visited so moving the remaining times closer together. What if we allow a 24-h clock?

Exercise 18.3 Suppose the ability to "pass" in chess is added. If both players "pass" then the game is drawn. Can we then prove that there is not an optimal strategy making the second player always win?

Exercise 18.4 There are n lions in a cage. A piece of meat is thrown in. Suppose

- The lions are hungry.
- A lion that eats the meat falls asleep.
- An asleep lion is meat to the other lions.
- Each lion and the meat is closest to exactly one other lion.
- The lions are ultra-intelligent.
- The lions prefer to stay alive.

What happens?

Chapter 19
Analytical Patterns

19.1 Introduction

When calculus is made rigorous, it is renamed analysis. It is one of the major branches
of modern pure mathematics. Analysis is essentially the study of limits. One of the big
achievements of 19th century mathematics was to give definite meaning to statements
such as

$$x_k \to x, \text{ as } k \to \infty,$$

and to derivatives. The reader who has done no analysis may find this chapter a little
tough, however, we try to present the basics in a slightly non-standard way that we
hope will be useful to both total beginners and novices.

The key to making sense of limits lay in a definition that appears slightly backwards
to the untrained eye. Before we proceed to it, we must define a sequence.

Definition 19.1 An $\mathbb{R}^n$−valued sequence is a map from $\mathbb{N}_1$ to $\mathbb{R}^n$

Thus we associate an element of $\mathbb{R}^n$ to every counting number, k. In analysis, we are
generally only concerned with what happens for k large. The truth of most statements
about limits will not change if the first one million values of the sequence are changed.
We can regard $\mathbb{C}$ as $\mathbb{R}^2$ so all our statements are equally valid for complex-valued
sequences.

In this chapter, we will look at the definition of limits and study some of the basic
patterns used to show that they do and do not exist. We can do no more than touch
on the basic concepts and ideas in a short chapter, however.

© Springer International Publishing Switzerland 2015
M. Joshi, *Proof Patterns*, DOI 10.1007/978-3-319-16250-8_19

19.2 The Triangle Inequality

A fundamental tool in analysis is the *triangle inequality*. This states that for two vectors $x, y \in \mathbb{R}^n$,

$$||x + y|| \leqslant ||x|| + ||y||.$$

The same result holds for $\mathbb{R}$ and $\mathbb{C}$, if we replace $||$ with $|$. It is often useful in a backwards way.

$$||v + w|| \geqslant ||v|| - ||w||.$$

To see this let $x = v + w$, $y = -w$.

19.3 The Definition

The modern definition of a limit of a sequence is as follows. We shall say that $x_k \to x$ as $k \to \infty$, if given any $\epsilon > 0$, there exists N such that for all $k \geqslant N$,

$$||x_k - x|| < \epsilon.$$

In other words, x_k tends to x if and only if you can pick any positive distance, wait long enough, and then eventually all points in the sequence are within that distance of x.

Sometimes N is written as N_ϵ to emphasize the fact that the value N will vary with the choice of ϵ. The definition does not say anything about the rate at which x_k gets close to x. It may be very fast or very slow.

The great feature of the definition is that it only involves finite quantities even though it is a statement about what happens as k gets very large and in some sense close to infinity. It is this feature that allowed the rigorization of calculus.

We now look at some examples. Let

$$x_k = \frac{1}{k}.$$

Intuitively, we would expect $x_k \to 0$. How do we show this with our definition? Given $\epsilon > 0$, let

$$N_\epsilon = 1 + \frac{1}{\epsilon}.$$

We then have that if $k \geqslant N_\epsilon$, then $k > 1/\epsilon$ which immediately implies

$$0 < \frac{1}{k} < \epsilon.$$

So the limit is zero as we would expect.

Having made the definition, we need to check that it is a good one. Our intuitive notion of limit suggests that a sequence should not have more than one limit. We prove that this is true with our definition.

Proposition 19.1 *Let* (x_k) *be a sequence in* $\mathbb{R}^n$. *Suppose* $x_k \to x$, *and* $x_k \to y$ *then* $x = y$.

Proof We show that $x = y$ by proving that they are within ϵ of each other for every $\epsilon > 0$. The only way this can happen is if they are equal. This is a common pattern in analysis. We call it *arbitrary closeness*.

Let $\epsilon > 0$. Let $\epsilon' = \epsilon/2$, then there exists integers $N_{\epsilon'}$, $M_{\epsilon'}$ such that

$$k > N_{\epsilon'} \implies ||x_k - x|| < \epsilon',$$
$$l > M_{\epsilon'} \implies ||x_l - y|| < \epsilon'.$$

Let $r = 1 + \max(N_{\epsilon'}, M_{\epsilon'})$. We then have

$$||x - y|| = ||x - x_r + x_r - y|| \leqslant ||x - x_r|| + ||x_r - y|| < 2\epsilon' = \epsilon.$$

So $x = y$ as claimed. □

Do all sequences have a limit? No! A simple example is

$$x_k = (-1)^k.$$

This alternates between -1 and 1 and so converges to neither. A more violent example is

$$y_k = k(-1)^k.$$

With our definition, we do not have a concept of converging to infinity. So a simple sequence that does not converge is

$$z_k = k.$$

However, one could make a different definition that encompasses the notion of converging to infinity and then it might converge. We do not explore that option here.

One simple property of limits is that they preserve weak inequalities.

Proposition 19.2 *Let* x_k *be a sequence in* $\mathbb{R}$. *Suppose* $x_k \to x$, *and* $x_k \geqslant y$ *for all* k *then* $x \geqslant y$. *Similarly for* $\leqslant$.

Proof We show that if $z < y$ then x_k does not converge to y. Let $\epsilon = y - z$. Then

$$|x_k - z| = x_k - z = (x_k - y) + (y - z) \geqslant y - z = \epsilon.$$

So there does not exist N such that for $k > N$, $|x_k - z| < \epsilon$. □

Strict inequalities are not preserved, however. We have

$$\frac{1}{k} > 0, \forall k \text{ and } \lim_{k \to \infty} \frac{1}{k} = 0.$$

19.4 Basic Results

It is fairly rare that the definition is used directly. Instead, results about convergence are used to imply that sequences converge from the fact that other sequences do. In our proof of the uniqueness of limits, we used a common technique which is that if we can show that convergence occurs within some multiple of ϵ that is enough. In fact, we can prove something better: *the poly-epsilon pattern*.

Proposition 19.3 *Suppose p is a non-zero polynomial with $p(0) = 0$, and all non-constant coefficients positive, (x_k) is a sequence in $\mathbb{R}^n$, $x \in \mathbb{R}^k$, $z > 0$ and for all $\epsilon > 0$, there exists N_ϵ such that $k > N_\epsilon$, implies*

$$||x_k - x|| \leqslant p(\epsilon^z),$$

then

$$x_k \to x.$$

As a special case, taking $z = 1$, and $p(y) = 2y$, we see that it is enough to show that there exists N_ϵ such that for $k > N_\epsilon$

$$||x_k - x|| \leqslant 2\epsilon.$$

Proof Suppose we are given $\epsilon > 0$. We know that given $f > 0$, we have for $k > N_f$,

$$||x_k - x|| \leqslant \sum_{j=1}^{d} c_j f^{jz}$$

where the polynomial is of degree d and

$$p(y) = \sum_{j=1}^{d} c_j y^j.$$

There is no constant term since $p(0) = 0$. Our objective is to pick e so that $k > N_e$ implies $||x_k - x|| < \epsilon$. Let $C = \max(c_j)$. We can write

$$||x_k - x|| \leqslant C \sum_{j=1}^{d} f^{jz}.$$

It is sufficient to consider the case where $\epsilon < 1$, since if we can show that the result holds with $\epsilon = 0.5$, then the same value of N will do for all $\epsilon > 0.5$. We therefore also take $f < 1$, the value of f^{jz} is biggest when $j = 1$, since raising any number between zero and one to a positive integer power makes it smaller. We therefore have

$$||x_k - x|| \leqslant C d f^z.$$

Now

$$C d f^z < \epsilon$$

if and only if

$$f^z < \frac{\epsilon}{Cd}.$$

This holds if and only if

$$f < \left(\frac{\epsilon}{Cd}\right)^{1/z}.$$

We therefore set

$$f = \frac{1}{2} \left(\frac{\epsilon}{Cd}\right)^{1/z},$$

and we have that for $k > N_f$

$$||x_k - x|| < \epsilon.$$

We have proven convergence.   □

With this result proven, we can prove some basic results.

Proposition 19.4 *Let $x_k \to x$ and let $y_k \to y$, then*

$$x_k + y_k \to x + y.$$

Proof Let $\epsilon > 0$, then there exists M, N such that

$$||x_k - x|| < \epsilon \quad \text{for} \quad k > M,$$

and

$$||y_k - y|| < \epsilon \quad \text{for} \quad k > N.$$

Let $L = \max(M, N)$. We then have for $k > L$,

$$||x_k + y_k - (x + y)|| = ||(x_k - x) + (y_k - y)|| \leqslant ||x_k - x|| + ||y_k - y|| < 2\epsilon.$$

Our result is now immediate by poly-epsilon.   □

Proposition 19.5 *Let $x_k \to x$ and let $\lambda \in \mathbb{R}$, then*

$$\lambda x_k \to \lambda x.$$

Proof Let $\epsilon > 0$, then there exists M such that

$$||x_k - x|| < \epsilon \quad \text{for} \quad k > M.$$

We have

$$||\lambda x_k - \lambda x|| = ||\lambda(x_k - x)|| = |\lambda|||x_k - x|| \leqslant |\lambda|\epsilon.$$

Our result is now immediate by poly-epsilon.   □

A slightly more complicated fact is that products of sequences converge to the product of the limits.

Proposition 19.6 *If $x_k \to x$, $y_k \to y$, in $\mathbb{R}$ or $\mathbb{C}$ then*

$$x_k y_k \to xy.$$

Proof Note that a sequence z_k converges to z if and only if $z_k - z$ converges to zero. We show that it is enough to consider the case where $x = y = 0$. To this observe

$$x_k y_k - xy = (x_k - x)(y_k - y) + xy_k + x_k y - 2xy.$$

The sum of the two middle terms converges to $2xy$ by our results above, so the sum of all four terms will converge to zero if and only if

$$(x_k - x)(y_k - y)$$

converges to zero. Given $\epsilon > 0$, we have that there exists M, N such that for $k > M, k > N$, $||x_k - x|| < \epsilon$, and $||y_k - y|| < \epsilon$. So for $k > L = \max(M, N)$,

$$||(x_k - x)(y_k - y)|| = ||x_k - x||||y_k - y|| < \epsilon^2.$$

The convergence follows by poly-epsilon.   □

Having established that sequences behave well under addition and multiplication, the obvious next question is what about subtraction and division? In fact, our existing results already cover subtraction. We have showed that if $x_k \to x$ then $-x_k \to -x$ by setting $\lambda = -1$ in the result above. So if $y_k \to y$, we have adding together that

$$x_k - y_k \to x - y.$$

This leaves division. We first prove a result about reciprocals.

Proposition 19.7 *If $x_k \to x$ in $\mathbb{R}$ or $\mathbb{C}$ and $x_k \neq 0$ for all k and $x \neq 0$, then*

$$\frac{1}{x_k} \to \frac{1}{x}.$$

Proof First note,

$$\left| \frac{1}{x_k} - \frac{1}{x} \right| = \left| \frac{x - x_k}{x x_k} \right|.$$

Dealing with the top is easy, the bottom is more subtle. We need to show that it is not too small so that its reciprocal is not too big. Taking $\epsilon = |x|/2$, we know that there exists N such that for $k > N$,

$$|x_k - x| < \frac{|x|}{2}.$$

This implies

$$|x_k| > \frac{|x|}{2}.$$

We thus have for $k > N$,

$$\left| \frac{1}{x_k} - \frac{1}{x} \right| < |x - x_k| \frac{2}{|x|^2}.$$

So given $\epsilon > 0$, we pick M_ϵ so that $M_\epsilon \geqslant N$, and $k > M_\epsilon$ implies

$$|x_n - x| < \epsilon.$$

So for $k > M_\epsilon$,

$$\left| \frac{1}{x_k} - \frac{1}{x} \right| < \frac{2}{|x|^2} \epsilon.$$

The result now follows by poly-epsilon.   □

Combining this result with our results on products, we have

Proposition 19.8 *If $x_k \to x$ in $\mathbb{R}$ or $\mathbb{C}$ and $x_k \neq 0$ for all k and $x \neq 0$, then if $y_n \to y$, we have*

$$\frac{y_k}{x_k} \to \frac{y}{x}.$$

We can now establish the convergence of various sequences. For example, if A and B are complex numbers

$$\frac{A}{k} \to A \times 0 = 0,$$

so

$$1 + \frac{A}{k} \to 1$$

and then

$$\frac{k+A}{k+B} = \frac{1+A/k}{1+B/k} \to 1.$$

Often when proving convergence, we use sandwiching or domination. The idea here is that if we have three sequences with one in between the other two, then the limiting behaviour of the middle must be between that of the other two.

Proposition 19.9 *Suppose x_k, y_k, z_k are real-valued sequences with*

$$x_k \leqslant y_k \leqslant z_k$$

for all k. If x_k and z_k converge to α then so does y_k.

Proof Let $\epsilon > 0$, there exist N_ϵ such that for $k > N_\epsilon$,

$$|x_k - \alpha| < \epsilon.$$

So

$$-\alpha - \epsilon < x_k \leqslant y_k.$$

Similarly, there exists M_ϵ such that for $k > M_\epsilon$,

$$\alpha + \epsilon > z_k \geqslant y_k.$$

So for $k > \max(L_\epsilon, M_\epsilon)$,

$$-\alpha - \epsilon < y_k < \alpha + \epsilon,$$

which is equivalent to

$$|y_k - \alpha| < \epsilon.$$

We are done.   □

For example, if $\beta > 1$, we immediately have

$$0 < \frac{1}{k^\beta} \leqslant \frac{1}{k}$$

so

$$\frac{1}{k^\beta} \to 0,$$

as $k \to \infty$.

When working with sequences in $\mathbb{R}^n$, it is sometimes more convenient to look at each coordinate. We can prove

Proposition 19.10 *The sequence x_k in $\mathbb{R}^n$ converges to $y \in \mathbb{R}^n$ if and only if every coordinate $x_{j,k}$ of x_k converges to the j coordinate of y.*

Proof Let $y = (y_1, \ldots, y_n)$. First, observe that

$$|x_{j,k} - y_j| \leqslant ||x_k - y||$$

so the forward direction is immediate.

Now suppose $x_{j,k} \to y_j$ for each j. Given $\epsilon > 0$, we can find $N_{j,\epsilon}$ for each j such that for $k > N_{j,\epsilon}$,

$$|x_{j,k} - y_j| < \epsilon.$$

Setting $N_\epsilon = \max_j N_{j,\epsilon}$, this holds for $j \geq N_\epsilon$.

Now

$$\begin{aligned}
||x_k - y|| &= \sqrt{\sum_{j=1}^n |x_{j,k} - y_j|^2} \\
&\leqslant \sqrt{n\epsilon^2}, \\
&= \sqrt{n}\epsilon.
\end{aligned}$$

Our result follows by poly-epsilon. $\qquad\square$

19.5 Series

Series are closely related to sequences but are often more interesting. With a series, it is the convergence of the partial sums that interests us not the sequence itself.

Definition 19.2 Let x_k be a real- or complex-valued sequence. Let

$$s_k = \sum_{i=1}^{k} x_i.$$

The series $\sum x_k$ is said to converge to s if and only if s_k converges to s. We call s_k the sequence of partial sums of x_i.

An important result when proving convergence of series is that a bounded increasing sequence of real numbers converges. We discussed this result in Chap. 10. Here we shall take it as an axiom.

This allows us to prove that series converge via domination.

Proposition 19.11 *Suppose $\sum x_k$ converges, and $0 \leqslant y_k \leqslant x_k$ for all k then $\sum y_k$ converges.*

Proof Since the terms are non-negative, both sequences of partial sums are increasing. We have that $\sum_{i=1}^{k} x_i$ is increasing and convergent to some limit, x. This implies

$$\sum_{i=1}^{k} y_i \leqslant \sum_{i=1}^{k} x_i \leqslant x$$

for all k. The sequence of partial sums of y_i is therefore increasing and bounded. The series converges. $\qquad\square$

For example, does $\sum k^\alpha$ converge? Before examining individual cases, we show that there exists α_0 such that for $\alpha < \alpha_0$ it converges and for $\alpha > \alpha_0$ it does not.

We have if $\alpha < \beta, k > 1$

$$k^\alpha < k^\beta.$$

So by domination, if $\sum k^\beta$ exists so does $\sum k^\alpha$. Contrapositively, if $\sum k^\alpha$ does not converge, neither does $\sum k^\beta$. Once we have found a value for which convergence occurs, all lower values give convergence. We can let α_0 be the supremum of the set of values for which convergence occurs.

If $\alpha = 0$, $\sum k^\alpha = \sum 1$ and divergence is clear. If we take $\alpha = -2$, we have for $k > 1$,

$$\frac{1}{i^2} < \frac{1}{i(i-1)} = \frac{1}{i-1} - \frac{1}{i}.$$

This means that

$$\sum_{i=1}^{k} i^{-2} \leqslant 1 + \sum_{i=2}^{k} \left(\frac{1}{i-1} - \frac{1}{i} \right) = 1 + 1 - \frac{1}{k} < 2.$$

So by domination

$$\sum_{i=1}^{k} i^{-2},$$

converges and $\sum_{i=1}^{k} i^{\beta}$ converges for all $\beta < -2$.

The changeover point is therefore between 0 and -2. We now look at -1. In this case, we have divergence. We will show that

$$s_{2^k} \geq 1 + \frac{k}{2}$$

which shows that the partial sums are unbounded and so convergence cannot occur.

Observe that we can write

$$s_1 = 1,$$

$$s_2 = 1 + \frac{1}{2},$$

$$s_4 = 1 + \frac{1}{2} + \frac{1}{3} + \frac{1}{4} > 1 + \frac{1}{2} + \frac{2}{4} = 1 + \frac{2}{2},$$

$$s_8 = s_4 + \frac{1}{5} + \frac{1}{6} + \frac{1}{7} + \frac{1}{8} > s_4 + \frac{4}{8} > 1 + \frac{2}{2} + \frac{1}{2}.$$

So we have shown the result for $k = 1, 2, 3, 4$. For the general case, we use induction, assume it is true for S_{2^k}, we then have

$$s_{2^{k+1}} = s_{2^k} + \sum_{l=2^k+1}^{2^{k+1}} \frac{1}{l} > s_{2^k} + (2^{k+1} - 2^k)2^{-(k+1)}.$$

The second term simplifies to $1/2$. So using the inductive hypothesis,

$$s_{2^{k+1}} > 1 + \frac{k}{2} + \frac{1}{2} = 1 + \frac{k+1}{2}$$

as required and we are done.

The series $1/k$ is sometimes called the *harmonic series*. We have seen that it diverges, however, the divergence is very slow. Our proof only shows that the sum of the first 2^{20} terms is at least 11, and 2^{20} is bigger than a million. It is an important example of a series in which the individual terms go to zero but the partial sums of the series go to infinity.

We have not addressed the convergence of k^α for $\alpha \in (-2, -1)$. In fact, these all converge. The standard way to proceed is to observe that

$$k^\alpha \leqslant x^\alpha$$

for $x \in (k - 1, k]$. It then follows that

$$\sum_{k=2}^{m} k^\alpha < \int_{1}^{m} x^\alpha dx = \frac{1}{1 + \alpha} \left(m^{1+\alpha} - 1 \right).$$

Showing that the integral is bounded is easy and the result follows. However, to do this properly would require us to develop a rigorous theory of integration which is beyond our scope.

We have seen that the terms of a series converging to zero does not imply that the series converges. However, the converse is true.

Proposition 19.12 If $\sum x_k$ exists then $x_k \to 0$.

Proof If the sum converges to x then given $\epsilon > 0$, there exists N such that for $n > N$

$$|s_n - x| < \epsilon.$$

Now

$$|s_{n+1} - s_n| \leqslant |s_{n+1} - x| + |x - s_n| < 2\epsilon.$$

But $x_{n+1} = s_{n+1} - s_n$. So for $k > N + 1$,

$$|x_k| < 2\epsilon,$$

and we are done.   □

This result is more often used in the contrapositive: if x_k does not converge to zero then $\sum x_k$ does not exist.

19.6 Continuity

Functions are a very general concept in mathematics. In some ways, they are too general to say much. A natural restriction is to require a function to be continuous.

Definition 19.3 Let $I \subset \mathbb{R}^n$. A function, x, from I to $\mathbb{R}^m$ is continuous if for every sequence x_k such that $x_k \in I$ for all k and $x_k \to x \in I$, we have

$$f(x_k) \to f(x).$$

In other words, we define continuity to mean that you can take the function through limits:
$$\lim_{k \to \infty} f(x_k) = f(\lim_{k \to \infty} x_k).$$

We have only considered real functions here but we can identify $\mathbb{C}$ with $\mathbb{R}^2$ so the definition and results work equally well for complex numbers. Note that the definition requires that f can be passed through the limit for *every* convergent sequence not just one of them.

What are some continuous functions? The identity function
$$f(z) = z,$$

is trivially continuous from our definition. We showed above that the limits of two convergent sequences is the product of the limits, applying this to the product of a sequence with itself, we have that
$$f(z) = z^2$$

is also continuous.

More generally, we can show

Proposition 19.13 *Suppose f, g are real (or complex-valued) continuous functions on I and $\lambda \in \mathbb{R}$ (or $\mathbb{C}$) then the following functions are continuous*

- $f + g$,
- λf,
- fg,

and if $g \neq 0$ everywhere and takes values in $\mathbb{R}$ (or $\mathbb{C}$) then so is f/g.

Proof If $x_n \to x$, then we have
$$f(x_n) \to f(x) \text{ and } g(x_n) \to g(x),$$

by the definition of continuity. Our results on limits then says
$$f(x_n) + g(x_n) \to f(x) + g(x).$$

This shows that $f + g$ is continuous. The proofs for λf and fg are essentially the same. For f/g we have that $g(x_n) \neq 0$ and $g(x) \neq 0$ so the result on quotients also goes over. □

We can now build up some continuous functions. If p is a polynomial then it is continuous on both $\mathbb{R}$ and $\mathbb{C}$. To see this, first we have by induction that z^k is continuous for all $k \in \mathbb{N}$. We also that $c_k z^k$ is continuous for any constant c_k. Summing a finite number of such terms yields all polynomials so all polynomials are continuous. If a polynomial, q, has no zeros in the set I then we also have that p/q is also continuous on I for any polynomial p.

There are plenty of continuous functions which are not polynomials. For example, let

$$f(x) = x^{1/2}$$

as a function from $[0, \infty)$ to $\mathbb{R}$. To show that it is continuous, we need to show that

$$x_k \to x \implies x_k^{1/2} \to x^{1/2}.$$

We use case analysis: either $x = 0$ or $x > 0$. If $x = 0$, we have to show that

$$x_k \to 0 \implies x_k^{1/2} \to 0.$$

Now given $\epsilon > 0$, there exists N such that $k > N$ implies

$$|x_k| < \epsilon.$$

and, so

$$|x_k^{1/2}| < \epsilon^{1/2}.$$

The result follows by poly-epsilon. If $x > 0$, then we have some N such that for $k > N$,

$$|x_k - x| < \frac{|x|}{2}$$

and so

$$|x_k| = |x_k - x + x| > \frac{1}{2}|x|.$$

This implies

$$|x_k|^{1/2} > \frac{1}{2}|x|^{1/2}.$$

We then have

$$|x_k^{1/2} - x^{1/2}| = \left| \left(x_k^{1/2} - x^{1/2} \right) \frac{x_k^{1/2} + x^{1/2}}{x_k^{1/2} + x^{1/2}} \right| = \frac{|x_k - x|}{|x_k^{1/2} + x^{1/2}|} < |x_k - x| \frac{2}{|x|^{1/2}}.$$

Using poly-epsilon, the result is now clear.

More generally, if we take the composition of two continuous functions, we get a continuous function. We simply pass each continuous function through the limit. This implies that $f(z) = |z|$ is a continuous function from $\mathbb{C}$ to $\mathbb{R}$ simply by taking the composition of

$$g(x + iy) = x^2 + y^2$$

and the square root function.

19.7 Theorems About Continuous Functions

The main reason to place a restriction on a class of objects is to make them have nice features. So what does continuity buy us? Our first result is that continuous functions have the property that they never jump in value. If a real-valued continuous function, f, on the reals takes the values y_0 at x_0 and the value y_1 at x_1, then all values between y_0 and y_1 are taken on $[x_0, x_1]$. This is sometimes called the *Intermediate Value Property*.

More formally, we want to show

Theorem 19.1 *Suppose $f : [x_0, x_1] \to \mathbb{R}$ is continuous and $f(x_0) < y < f(x_1)$ then there exists $x \in (x_0, x_1)$ such that $f(x) = y$.*

Proof We construct a convergent sequence whose limit is the required point x. In fact, we construct two sequences. Let $a_0 = x_0$ and let $b_0 = y_0$. Now consider $c_0 = 0.5(a_0 + b_0)$. If $f(c_0) = y$ we are done. If $f(c_0) < y$, let $a_1 = c_0, b_1 = b_0$. Otherwise, let $a_1 = a_0, b_1 = c_0$.

We now have the same situation as before with the endpoints $[a_1, b_1]$. And $b_1 - a_1 = 0.5(b_0 - a_0)$. We now simply repeat and stop if for j $f(c_j) = y$. If this never happens we have sequences (a_j) and (b_j) such that

$$b_j - a_j = 2^{-j}(b_0 - a_0),$$

$$f(a_j) < y < f(b_j).$$

The sequence a_j is increasing and bounded above by b_0 so it converges to some point x. Now

$$f(x) = f(\lim_j a_j) = \lim_j f(a_j) \leqslant y$$

since $f(a_j) \leqslant y$ for all j. Now b_j will also converge to x since

$$|b_j - x| = b_j - x < b_j - a_j = 2^{-n}|b_0 - a_0|.$$

So

$$f(x) = f(\lim_j b_j) = \lim_j f(b_j) \geqslant y,$$

since $f(b_j) > y$ for all j. We have proven

$$f(x) = y$$

and we are done.   □

Note that the crucial feature of f in this proof was that we could pass limits through it which was precisely our definition of continuity.

A second important property of continuous functions is that they always have a maximum and a minimum on closed and bounded intervals. Note that is not true without all three of these hypotheses, for example, consider

$$f(x) = 1/x$$

on the set $(0, 1)$. The image set is $(1, \infty)$. At no point is a minimum value attained and the maximum is infinity. Our theorem is

Theorem 19.2 *Let $f : [a, b] \to \mathbb{R}$ be continuous then there exist $x, y \in [a, b]$ such that*

$$f(x) \leqslant f(s) \leqslant f(y)$$

for all $s \in [a, b]$.

Proof We can deduce the result for the minimum from that for the maximum by considering the function $-f$, so it is enough to consider the maximum. We again use a two-sequence argument. Let $a_0 = a, b_0 = b$. Now let

$$c_0 = 0.5(a_0 + b_0).$$

We shall say that f is at least as big on $[a_0, c_0]$ as on $[c_0, b_0]$ if for all $x \in [c_0, b_0]$, there is a point x' in $[a_0, c_0]$ such that

$$f(x) \leqslant f(x').$$

If this is the case we let $a_1 = c_0, b_1 = b_0$, otherwise we let $a_1 = a_0, b_1 = c_0$. In the second case, there is a point $x' \in [a_1, b_1]$ such that

$$f(x') > f(x)$$

for all $x \in [a_0, c_0]$. The crucial fact is now that for each $s \in [a_0, b_0]$, there exists a point $t \in [a_1, b_1]$ where

$$f(s) \leqslant f(t).$$

If a maximum exists it must therefore be in $[a_1, b_1]$.

We can now repeat. We get intervals $[a_j, b_j]$ and these retain the property that for each point in $[a_0, b_0]$ there is some point in $[a_j, b_j]$ where f is at least as big.

As for the intermediate value theorem, the sequences a_j and b_j converge to the same point x. It is at this point where the maximum occurs. We still have to prove this, however.

We know that for each $s \in [a, b]$ and $j \in \mathbb{N}_1$, there exists a point $s_j \in [a_j, b_j]$ with $f(s) \leqslant f(s_j)$ by our construction of the intervals $[a_j, b_j]$. This sequence s_j will also converge to x by the "sandwich theorem." We have

$$f(s) \leqslant f(s_j)$$

for all j. We have

$$f(x) = \lim f(s_j) \geqslant f(s).$$

So x is indeed a maximum for f on $[a, b]$ as claimed. $\qquad\square$

Note that we did not show that the maximum is unique, and, in general, it will not be. For example, if the function f is constant every point in $[a, b]$ is a maximum.

An analogue of this result holds true for continuous functions on $\mathbb{R}^n$. The crucial feature of the set I above is that it is closed and bounded.

Theorem 19.3 *Let* $E \subset \mathbb{R}^n$ *be a product of intervals* $I_j = [\alpha_j, \beta_j]$ *and let* $f : E \to \mathbb{R}$ *be continuous, then* f *has a maximum and a minimum in* E.

Proof We can do the same proof as for the one-dimensional case. We simply divide each I_j into two equal sub-intervals. This divides E into 2^n pieces. Given any two pieces, one of them must have the property that f is bigger on it in the same sense as the proof for the one dimensional case. So repeatedly comparing the 2^n pieces, we find one which has that property compared to all the others. We now simply repeat the exercise on that piece and get a sequences of sets E_k with $E_k \subset E_{k-1}$ and E_k is a product of intervals, $[\alpha_{j,k}, \beta_{j,k}]$, with the length of the kth coordinate interval being $2^{-k}(\beta_j - \alpha_j)$. For every point in E, there is a point in E_k where f is at least as big by construction.

By the same arguments as for the one-dimensional case, there exists x_j such that

$$\lim_k \alpha_{j,k} = x_j = \lim_{j,k} \beta_{j,k}.$$

The points $x = (x_1, x_2, \ldots, x_n)$ is the limit of the sequences

$$(\alpha_{1,k}, \alpha_{2,k}, \ldots, \alpha_{n,k})$$

and it is the maximum for the same reasons as before. $\square$

Whilst we have proven the result for cuboids, the product of intervals structure is certainly not necessary. For example, if we take any finite union of sets for which it holds, then it will still hold; simply find the maximum on each, and then take the maximum on that small set of points. One can also prove the result for balls of the form, $\|x\| \leqslant R$, by a similar argument that is just more fiddly since the endpoints are curved.

19.8 The Fundamental Theorem of Algebra

As a culminating theorem in this book, we consider the proof of the theorem that every complex polynomial has a complex root: this is sometimes called the *fundamental theorem of algebra*. We adapt a proof from "Proofs from the BOOK." We first show a lemma sometimes called d'Alembert's Lemma or d'Argand's inequality.

Lemma 19.1 *Suppose p is a complex polynomial. Suppose z_0 is not a zero of p, then for any $\epsilon > 0$, there exists a point z such that*

$$|z - z_0| < \epsilon \text{ and } |p(z)| < |p(z_0)|.$$

Proof Any point within ϵ of z_0, can be written as

$$z_0 + w$$

with $|w| < \epsilon$. Let p be of degree d. Since p is a polynomial, we can consider the polynomial

$$q(w) = p(z_0 + w) - p(z_0)$$

as a polynomial in w with z_0 fixed. Clearly, $q(0) = 0$ that is 0 is a root of q. We can therefore write

$$q(w) = wq_1(w)$$

with q_1 a polynomial of one lower degree. Either $q_1(0) \neq 0$, or we can write $q_1(w) = wq_2(w)$. Repeating, we obtain

$$q(w) = w^k q_k,$$

where q_k is of degree $d - k$ and $q_k(0) \neq 0$. Note that this process must stop when $k = d$, if not before, or p is identically zero.

We thus have that we can write

$$p(w + z_0) = p(z_0) + w^k q_k(w),$$

with $q_k(0) \neq 0$.

The idea of the final part of the proof is to find a direction which makes $w^k q_k(w)$ point in the opposite direction to $p(z_0)$. That is we make it a positive real multiple of the negative of $p(z_0)$. This will be sufficient since it will add to $p(z_0)$ to give a smaller number provided its magnitude is smaller. Sufficiently close to $w = 0$, this will be true as we have a factor of w^k.

Since we can write

$$p(z_0) = Re^{i\theta}$$

with R and θ real, a complex number z will be a real multiple of $-p(z_0)$ if it is of the form

$$z = Se^{i(\theta + \pi)}$$

with S real.

We can write

$$q_k(0) = Pe^{i\phi}$$

with P, ϕ real and $P > 0$.

As a warm-up, we first do the special case that q_k is constant. We want to find γ such that if $w = e^{i\gamma}$, then

$$(e^{i\gamma})^k e^{i\phi} = e^{i(\theta + \pi)}.$$

We thus need

$$k\gamma + \phi = \theta + \pi$$

up to a multiple of 2π. We simply set

$$\gamma = \frac{1}{k}(\theta + \pi - \phi),$$

and the q_k constant case is done.

In general, of course, q_k will not be constant. However, we can write

$$q_k(w) = q_k(0) + wr(w)$$

for a polynomial r of lower degree. We will make the term $wr(w)$ too small to affect our results. In particular, we know that r is continuous so $|r|$ is bounded by some value M on the set we are interested in. We therefore have that $|wr(w)|$ is less than $M|w|$. Take γ as in the constant case, and let $w = Se^{i\gamma}$ with S real. We then have

$$p(z_0 + w) = Re^{i\theta} + S^k P e^{ik\gamma + i\phi} + w^{k+1} r(w).$$

By our choice of γ we now have

$$p(z_0 + w) = (R - PS^k)e^{i\theta} + w^{k+1} r(w).$$

So for S small the first term is certainly of smaller modulus than $p(z_0)$. For the second term, we can bound it by

$$MS^{k+1}.$$

So

$$|p(z_0 + w)| < |R - PS^k| + MS^{k+1}$$

by the triangle inequality. This equals

$$R - PS^k + MS^{k+1}$$

for S small. Since S^{k+1} goes to zero faster than S^k, for S sufficiently small, this is less than $R = |p(z_0)|$, and we are done. $\square$

The lemma is important in that it shows that if $p(z_0)$ is non-zero then z_0 is not a local minimum of $|p(z)|$ and therefore it is not a global minimum either. In other words, a polynomial p has to be zero at a global minimum of $|p|$. So to prove that a zero exists all we have to do is show that a global minimum of $|p|$ exists. However, this is easy. We can write

$$p(z) = \sum_{i=0}^{d} c_i z^i$$

for some complex numbers c_i with $c_d \neq 0$. So

$$c_d z^d = p(z) - \sum_{i=0}^{d-1} c_i z^i$$

and

$$|c_d||z|^d \leqslant |p(z)| + \left| \sum_{i=0}^{d-1} c_i z^i \right|,$$

$$\leqslant |p(z)| + \sum_{i=0}^{d-1} |c_i||z|^i.$$

This implies that

$$|p(z)| \geqslant |c_d||z|^d - \sum_{i=0}^{d-1} |c_i||z|^i.$$

The highest power $|z|^d$ will dominate on the right side for $|z|$ large. So as $|z|$ goes to infinity, the right hand side will go to infinity. Hence so will $|p(z)|$. In particular, we can find R such that everywhere on the circle $|z| = R$, $|p(z)| > |p(0)|$. Since p is continuous on the set $|z| \leqslant R$, then, as we saw above, it has a minimum in it. The minimum is not on the boundary since everywhere on the boundary $|p(z)|$ is bigger than $|p(0)|$. At this minimum, z_0, we must have $p(z_0) = 0$, since otherwise it would not be a minimum by the lemma. We are done.

19.9 Further Reading

Analysis is a huge field and there are myriads of books. My favourite introductory book is Byrant's "Yet Another Introduction to Analysis." This book is anything but yet another one on the topic! The author works hard to motivate everything he does and to keep everything clear and elementary. He keeps the topic rigorous but also concrete. As a follow-up, I recommend Binmore's "Mathematical Analysis: A Straightforward Approach" which again has an emphasis on simplicity and clarity.

19.10 Problems

Exercise 19.1 Prove or disprove that the following sequences converge. If they converge, identify the limit.

- $x_k = k$,
- $x_k = 1 - \frac{1}{k}$.
- $x_k = \frac{1+k}{2+k^2}$.
- $x_k = \sqrt{k+1} - \sqrt{k}$.

Exercise 19.2 Suppose $f : \mathbb{R}^2 \to \mathbb{R}$, is continuous on the set

$$E = \{1 \leqslant |(x, y)| \leqslant 2\},$$

show that f has a maximum and minimum on E. Find an example where these occur on the boundary.

Exercise 19.3 A function is said to have the intermediate value property on $\mathbb{R}$ if $f : \mathbb{R} \to \mathbb{R}$, and if $a < b$, $f(a) < x < f(b)$ implies that there exists $c \in (a, b)$ with $f(c) = b$. Let a function g have value 0 at 0 and value

$$g(x) = \sin\left(\frac{1}{x}\right)$$

for $x \neq 0$. Does g have the intermediate value property? Is g continuous?

Exercise 19.4 Does each of the following series converge? Prove or disprove.

- $x_k = \frac{1}{1+k}$.
- $x_k = \frac{1}{1+k^2}$.
- $x_k = \frac{1}{1+k^3}$.

Chapter 20
Counterexamples

20.1 Introduction

Developing new mathematics is as much about disproving conjectures as it is about proving theorems. The standard way to disprove a conjecture is to construct an example satisfying the proposed hypotheses but not the conclusion. Such an example is called a *counterexample*. The process of mathematics is then that the conjecture's hypotheses are modified to outlaw the counterexamples and then the mathematicians again try to prove or disprove. If the counterexample is in the spirit of the conjecture, they may well conclude that the direction is wrong and interest moves on to other topics. Alternatively, someone may well feel that there still is a positive result to be found, and it is simply a question of working out the correct hypotheses to rule out unnatural examples.

The big difference between proving theorems and constructing counterexamples is specificity. A counterexample is just a single solitary example that rules out a conjecture being true. This contrasts with proving a theorem where we are trying to prove many cases at once.

Knowing a good collection of examples and counterexamples is an important part of being a mathematician. Any ideas can be quickly tested against this collection and modified appropriately or dropped. In this chapter, we present some elementary examples that show that more hypotheses than one might guess are necessary to prove some basic results.

20.2 Matrix Algebra

Matrices are an important part of linear algebra and provide a good example of a collection of objects which have some of the properties of real numbers but not all of them. Recall that if we have two by two matrices then

© Springer International Publishing Switzerland 2015
M. Joshi, *Proof Patterns*, DOI 10.1007/978-3-319-16250-8_20

$$\begin{pmatrix} a_{11} & a_{12} \\ a_{21} & a_{22} \end{pmatrix} \cdot \begin{pmatrix} b_{11} & b_{12} \\ b_{21} & b_{22} \end{pmatrix} = \begin{pmatrix} a_{11}b_{11} + a_{12}b_{21} & a_{11}b_{12} + a_{12}b_{22} \\ a_{21}b_{11} + a_{22}b_{21} & a_{21}b_{12} + a_{22}b_{22} \end{pmatrix}.$$

One can easily conjecture that matrix multiplication has the same properties as multiplication of real numbers. We will construct some counterexamples to show that some of them do not hold.

To see that matrix multiplication is not commutative, in general, it is enough to construct one counterexample. So we have

Conjecture 20.1 *If A and B are square matrices then $AB = BA$.*

Disproof by counterexample

$$\begin{pmatrix} 1 & 1 \\ 1 & 1 \end{pmatrix} \begin{pmatrix} 0 & 1 \\ 0 & 0 \end{pmatrix} = \begin{pmatrix} 0 & 1 \\ 0 & 1 \end{pmatrix};$$

$$\begin{pmatrix} 0 & 1 \\ 0 & 0 \end{pmatrix} \begin{pmatrix} 1 & 1 \\ 1 & 1 \end{pmatrix} = \begin{pmatrix} 1 & 1 \\ 0 & 0 \end{pmatrix}.$$

In this case, we see that the product of the two matrices depends on order and thus matrix multiplication does not commute in general. Note that it is still possible for a given pair of matrices A and B that $AB = BA$ holds.

Now consider the question of the existence of inverses. To invert a matrix, A, we need to find a matrix B such that

$$AB = I$$

where I is the identity matrix:

$$I = \begin{pmatrix} 1 & 0 \\ 0 & 1 \end{pmatrix}.$$

Conjecture 20.2 *If A is a square matrix then there exist a matrix, B, such that*

$$AB = I.$$

Disproof by counterexample
Let

$$A = \begin{pmatrix} 1 & 0 \\ 0 & 0 \end{pmatrix}.$$

We multiply by a general matrix

$$\begin{pmatrix} 1 & 0 \\ 0 & 0 \end{pmatrix} \begin{pmatrix} a & b \\ c & d \end{pmatrix} = \begin{pmatrix} a & b \\ 0 & 0 \end{pmatrix}.$$

Whatever values we take for a, b, c and d, we do not obtain I. We conclude that A is not invertible. Note that in order to show that the particular example worked in this case, we did need to consider all possible inverses to show that none of them worked.

In ordinary arithmetic, the square of a non-zero number is never zero. What about matrices?

Conjecture 20.3 *If A is a non-zero square matrix then* $A^2 \neq 0$.

Disproof by counterexample

$$\begin{pmatrix} 0 & 1 \\ 0 & 0 \end{pmatrix}^2 = \begin{pmatrix} 0 & 0 \\ 0 & 0 \end{pmatrix}.$$

20.3 Smooth Functions

In this section, we present some counterexamples which show that some natural conjectures about smooth functions are, in fact, false.

Conjecture 20.4 *Suppose* $f : \mathbb{R} \to \mathbb{R}$ *is infinitely differentiable and defines a bijection and so has an inverse g. We conjecture that g must also be infinitely differentiable.*

Disproof by counterexample
Let f be the function with
$$f(x) = x^3.$$

The inverse function is then given by

$$g(x) = x^{1/3}.$$

The function is not differentiable at $x = 0$ and for $x \neq 0$, we have

$$g(x) = \frac{1}{3}x^{-2/3}$$

which blows up at 0.

Conjecture 20.5 *Suppose* $f : \mathbb{R} \to \mathbb{R}$ *is infinitely differentiable and defines a bijection. We conjecture that we must have for all x that* $f'(x) \neq 0$.

Disproof by counterexample
Again, let f be the function with

$$f(x) = x^3.$$

We have $f'(0) = 0$.

An important property of continuous functions is the *intermediate value property*. This says that if a function takes the value $f(a)$ at a and the value $f(b)$ at b then it takes all the values in between in the interval (a, b). More formally, it says that if $a < b$ and $f(a) < y < f(b)$ or $f(a) > y > f(b)$ then there exists $c \in (a, b)$ such that

$$f(c) = y.$$

It is a theorem that continuous functions have this property (Theorem 19.1). Indeed, one might take the property as a definition of continuity rather than the standard one. So the question becomes "Is every function that has the intermediate value property continuous?"

Conjecture 20.6 *Let* $f : \mathbb{R} \to \mathbb{R}$ *have the intermediate value property for every interval* $[a, b] \subset \mathbb{R}$, *then* f *is continuous.*

Disproof by counterexample
Let

$$f(x) = \sin(1/x)$$

for $x \neq 0$ and 0 at $x = 0$. The function f is a composition of two continuous functions away from 0 and so is continuous except at 0.

However, it has very odd behaviour at zero. In particular, if we take a sequence x_n such that x_n tends to zero, a continuous function should have

$$f(x_n) \to f(0) = 0.$$

However, if we let

$$x_n = \left(\frac{\pi}{2} + 2\pi n\right)^{-1},$$

we have $f(x_n) = 1$. We can construct similar sequences converging to any value in $[-1, 1]$.

However, f does have the intermediate value property. To see this, first observe that if 0 is not in an interval $[a, b]$ we are studying a continuous function so all its properties hold, so we need only consider the case $0 \in [a, b]$. We take $b > 0$ for simplicity. However, a very similar argument works for intervals $[a, 0]$. The function $\sin$ takes all values between -1 and 1 on any interval

$$[2\pi j, 2\pi(j + 1)]$$

for $0 \neq j \in \mathbb{N}$ and never takes any value outside $[-1, 1]$. This means that f takes these values on any interval

$$[1/(2\pi(j + 1)), 1/(2\pi j)].$$

For j sufficiently large, this will be within $[a, b]$ and so there must be a point in the interval for each value between -1 and 1 which is sufficient given that $f(a)$ and $f(b)$ must also be in that range.

20.4 Sequences

Much of analysis is dominated by questions involving interchanges of limiting operations. In other words, if we have two parameters does the order of taking them to limits matter? The simplest context in which to study these questions is sequences. A two-dimensional sequence is a map from $\mathbb{N} \times \mathbb{N}$ to $\mathbb{R}$. If we fix either coordinate, we get an ordinary sequence.

Conjecture 20.7 *Suppose we have a two-dimensional sequence $x_{m,n}$ such that for all m, n,*

$$\lim_{m \to \infty} x_{m,n} \text{ and } \lim_{n \to \infty} x_{m,n}$$

exist. Suppose also that

$$\lim_{n \to \infty} \lim_{m \to \infty} x_{m,n} \text{ and } \lim_{m \to \infty} \lim_{n \to \infty} x_{m,n}$$

exist. We conjecture that the last two limits are equal.

Disproof by counterexample
Let

$$x_{m,n} = \frac{m}{m + n}$$

for $m, n > 0$. We then have

$$\lim_{m \to \infty} x_{m,n} = \lim_{m \to \infty} \frac{m}{m + n} = \lim_{m \to \infty} \frac{1}{1 + n/m} = 1,$$

but

$$\lim_{n \to \infty} x_{m,n} = \lim_{n \to \infty} \frac{m}{m + n} = 0.$$

So the iterated limits are 1 and 0 and do not agree.

A related example regards the limits of integrals. Can we always take a limit through an integral?

Conjecture 20.8 *Suppose $f_n : [0, 1] \to \mathbb{R}$ is continuous for each n, and suppose for all $x \in [0, 1]$, $f_n(x) \to f(x)$ with f continuous. We conjecture that*

$$\lim_{n \to \infty} \int_0^1 f_n(x) dx = \int_0^1 f(x) dx.$$

Note that we have not said what sort of integral this is, e.g. Lebesgue or Riemann. However, for a continuous function on a finite interval all reasonable integrals give the same answer.

Disproof by counterexample
Let

$$f_n(x) = \begin{cases} n^2 x \text{ for } x \in [0, 1/n], \\ n^2(2/n - x) \text{ for } x \in (1/n, 2/n], \\ 0 \text{ for } x > 2/n. \end{cases} \tag{20.4.1}$$

A direct computation yields that

$$\int\limits_0^1 f_n(x)dx = 1$$

for every n. However, for any given x the value of $f_n(x)$ is eventually zero as n goes to infinity. So the limit function is 0 everywhere and has 0 integral.

20.5 Ordinary Differential Equations

It is normal in the study of ordinary differential equations to talk of "the solution" as if a solution always exists and is unique. We will see in this section that this is not always the case.

Conjecture 20.9 *Suppose that $f : \mathbb{R} \to \mathbb{R}$ is a continuous function, then there is a unique function y from $\mathbb{R}_+$ to $\mathbb{R}$ such that $y(0) = 0$ and*

$$\frac{dy}{dx} = f(y).$$

Disproof by counterexample
Let

$$f(z) = 3z^{2/3}.$$

An easy solution is simply $y(x) = 0$ for all x. However, there is also another solution. We let

$$y(x) = x^3.$$

We then certainly have that $y(0) = 0$ and that

$$\frac{dy}{dx} = 3x^2 = f(y).$$

This example shows that we need extra hypotheses to get a positive result regarding uniqueness. In fact, there is a standard additional condition which does lead to uniqueness. Generally, f is assumed to be assumed Lipschitz continuous that is there exists a constant C such that for all s and t

$$|f(s) - f(t)| \leqslant C|s - t|.$$

We do not present the proof here since we are focusing on counterexamples.

20.6 Characterisation

It is worth mentioning that counterexamples are not the only way to show that statements are not universally valid. An alternate approach is show that the desired property is equivalent to some simpler property. For example, if we take two by two matrices, we can use the determinant to characterise invertibility.

First, if the determinant is non-zero, there is a simple formula for the inverse. Let

$$A = \begin{pmatrix} a & b \\ c & d \end{pmatrix}.$$

The determinant of A is defined by

$$\det(A) = ad - bc.$$

It follows from direct computation that if $\det(A)$ is non-zero then

$$A^{-1} = \frac{1}{\det A} \begin{pmatrix} d & -b \\ -c & a \end{pmatrix}$$

is both a left and right inverse that is

$$A^{-1}A = AA^{-1} = I.$$

So we have that if $\det(A)$ is non-zero then A is invertible.

The next question is can a matrix with zero determinant be invertible? We show that it cannot. Let A and B be two by two matrices, then

$$\det(AB) = \det(A)\det(B).$$

This can be checked by direct computation. It is also true that

$$\det(I) = 1.$$

So if
$$AB = I$$

then
$$\det(A)\det(B) = 1.$$

So $\det(A)$ is non-zero.

We have shown that a matrix is invertible if and only if its determinant is non-zero. We therefore immediately see that there are many non-invertible matrices. For example, if $a \neq 0$ and we set
$$d = bc/a,$$

for any b and c.

Although we have only studied two by two matrices in this section, similar results hold for general square matrices. It is simply a question of obtaining the right definition for a determinant.

20.7 Problems

Exercise 20.1 Find matrices A and B which are distinct and $AB = BA \neq 0$.

Exercise 20.2 Find a matrix with all entries non-zero which is not invertible.

Exercise 20.3 Do there exist square matrices, A, with $A^2 = 0$ and all entries non-zero?

Exercise 20.4 If A is a square matrix and there exists B such that $AB = I$, must there exist C such that
$$CA = I,$$

and if C does exist, must we have $C = B$?

Exercise 20.5 Which of the following functions are Lipschitz continuous?

- $f(x) = x^{2/3}$.
- $f(x) = x$.

Exercise 20.6 Suppose $f : (0, 1) \to \mathbb{R}$ is continuous. Must there be a global maximum in $(0, 1)$? i.e. does there always exist x such $f(y) \leqslant f(x)$ for all $y \in (0, 1)$?

Appendix A
Glossary

In this appendix, we define some common terms and establish notation for the reader's convenience.

First, we define our number systems.

- The integers, $\mathbb{Z}$, are $\{\ldots, -3, -2, -1, 0, 1, 2, 3, \ldots\}$.
- The natural numbers, $\mathbb{N}$, are $\{0, 1, 2, 3, \ldots\}$. They include zero. They are the non-negative integers.
- The counting numbers, $\mathbb{N}_1$, are $\{1, 2, 3, \ldots\}$. They do not include zero. They are also known as the positive integers.
- The real numbers will be denoted $\mathbb{R}$.
- The complex numbers will be denoted $\mathbb{C}$.

Next, we define terminologies for mappings.

- The *domain* of a mapping from X to Y is X. The set Y is sometimes called the *codomain*.
- An *injection* is a map, f, between sets X and Y such that

$$x_1 \neq x_2 \implies f(x_1) \neq f(x_2).$$

Equivalently,

$$f(x_1) = f(x_2) \implies x_1 = x_2.$$

The mapping f is said to be *injective* or *one-one*.
- The *range* of a mapping from X to Y is the set

$$\{f(x), \ x \in X\}.$$

- A *surjection* is a map f between sets X and Y such that the range of X is Y. Such a map is said to be *surjective* or *onto*.

© Springer International Publishing Switzerland 2015
M. Joshi, *Proof Patterns*, DOI 10.1007/978-3-319-16250-8

We now define some terminology for sequences.

- A *sequence* is a mapping from $\mathbb{N}_1$ to a set Y. The set Y is generally $\mathbb{R}$ or $\mathbb{C}$. The mapping is generally written as (x_j) or $x_1, x_2, x_3, \ldots$ rather than as $f(1), f(2), f(3), \ldots$
- A sequence, (x_j), is *bounded* if there exists some $K \in \mathbb{R}$ such that

$$|x_j| \leq K$$

for all j.
- A sequence, (x_j), is *increasing* if for all j

$$x_j \leq x_{j+1}.$$

- A *series* is a sequence of partial sums of a sequence. Thus if (x_j) is a sequence, we have the series (S_n) defined by

$$S_n = \sum_{j=1}^{n} x_j.$$

Appendix B
Equivalence Relations

In this appendix we recall the basics of equivalence relations. First, recall that a *relation* on a set X is a subset, R, of $X \times X$. We say that x relates to y if

$$(x, y) \in R.$$

We shall write this as

$$x \sim y.$$

Such a relation is said to be *reflexive* if every element relates to itself so

$$x \sim x, \ \forall x \in X.$$

It is *symmetric* if

$$x \sim y \implies y \sim x.$$

In others words, $(x, y) \in R$ implies that $(y, x) \in R$. It is *transitive* if

$$x \sim y, y \sim z \implies x \sim z.$$

Given an element x of X we define the equivalence class of x written as $[x]$ to be the subset of X given by

$$\{y \in x \mid x \sim y\}.$$

An important property of equivalence classes is that

$$[x] = [y] \iff x \sim y.$$

First, we show that $x \sim y$ implies $[x] = [y]$. If $z \in [x]$ then $x \sim z$ so $z \sim x$ and $x \sim y$, which implies $z \sim y$ by transitivity. We have $y \sim z$ by symmetry, and

© Springer International Publishing Switzerland 2015
M. Joshi, *Proof Patterns*, DOI 10.1007/978-3-319-16250-8

we have shown $z \in [y]$. This means that $[x] \subseteq [y]$. However, swapping x and y everywhere also demonstrates $[y] \subseteq [x]$, and so by symmetry

$$[x] = [y].$$

For the opposite direction, if $[x] = [y]$, then $y \in [x]$ since $y \in [y]$ by reflexivity. Saying $y \in [x]$ is by definition $x \sim y$ and we are done.

Equivalence relations are important because they induce partitions of X. A *partition* of X is a decomposition into smaller sets X_j such that every element of X is in precisely one subset X_j. So

$$X = \bigcup X_j,$$

and for $j \neq k$

$$X_j \cap X_k = \varnothing.$$

We do not make any particular requirement on how many sets X_j there are.

Our main result is

Theorem B.1 *If R is an equivalence relation on X then the equivalence classes induced by R define a partition of X.*

Proof First, we have assumed that R is reflexive so every x is in some equivalence class, and the union of all the classes is indeed X. Second, we need to show that if two equivalence classes have non-empty intersection then they are equal. If

$$z \in [x] \cap [y]$$

then

$$x \sim z, \ y \sim z.$$

By symmetry, $z \sim y$. By transitivity, it then follows that $x \sim y$. We showed above that this implies

$$[x] = [y]$$

and we are done.   □

Alternatively, given a partition, X_j, we can define a relation by

$$x \sim y \iff \exists k, \ x, y \in X_k.$$

That is x relates to y if and only if they are in the same subset when the set is partitioned.

References

1. Aigner M, Ziegler GM (2012) Proofs from the book, 4th edn. Springer, Berlin
2. Binmore K (1983) Mathematical analysis: a straight-forward approach, 2nd edn. Cambridge University Press, Cambridge
3. Bryant V (1990) Yet another introduction to analysis. Cambridge University Press, Cambridge
4. Bold B (1982) Famous problems of geometry and how to solve them. Dover, New York
5. Eccles PJ (1997) An introduction to mathematical reasoning. Cambridge University Press, Cambridge
6. Enderton HB (1977) Elements of set theory. Academic Press, New York
7. Oeschslin W (2010) Elements of Euclid by Byrne. Taschen, Koln
8. Golomb SW (1956) A combinatorial proof of Fermat's "Little" theorem. Am Math Monthly 63(10):718. http://www.jstor.org/stable/2309563
9. Houston K (2009) How to think like a mathematician: a companion to undergraduate mathematics. Cambridge University Press, Cambridge
10. Gura E-Y, Maschler MB (2008) Insights into game theory. Cambridge University Press, Cambridge
11. Halmos PR (2014) Naive set theory, 2nd edn. Springer, New York
12. Velleman DJ (2006) How to prove it: a structured approach, 2nd edn. Cambridge University Press, Cambridge
13. Wilson R (2002) Four colours suffice: how the map problem was solved. Allen Lane Science, London

© Springer International Publishing Switzerland 2015
M. Joshi, *Proof Patterns*, DOI 10.1007/978-3-319-16250-8

Index

A
Abalone, 147
Addition modulo k, 60
Additive subgroup, 43
Aleph, 114
Algebraic number, 65, 70, 116
Algorithm, 21, 28
Algorithmic construction, 25, 28, 143
Analysis, 151
Appel
    Kenneth, 106
Arbitrary closeness, 153
Associative, 59, 60
Associativity, 84

B
Backwards strategy, 149
Base, 21
Basis, 55, 67
Bijection, 55, 58, 60, 68, 109
Binary number, 114
Binomial coefficient, 4
Binomial theorem, 5
Bold
    Benjamin, 80
Bounded, 120, 182

C
Calculus, 151
Canonical, 90
Cardinality, 58, 59, 61, 109, 121
Cartesian product, 60, 82
Case analysis, 105, 164
Characterisation, 179
Chess, 147, 149

Chess-board, 53
Classic game, 147
Classification
    proof by, 97
Closed, 44, 122
Closure, 46
Codomain, 181
Commutative, 59, 60, 84
Commutative group, 59, 60
Complete induction, 1, 28, 101
    proof of, 3
Complex number, 65, 81, 115, 116
Component, 120, 121
Composite number, 2
Composition of continuous functions, 165
Connected, 106, 119, 130
Continuous, 58
Continuous function, 162, 165, 167
Contradiction
    proof by, 33–41, 147
Contrapositive, 101, 162
    proof by, 33–41
Convergence by domination, 39
Convergent sequence, 163
Converse, 40
Convex, 44, 120, 129
Convex hull, 46
Convex polyhedron, 129
Convex set
    connectedness of, 119
Countable, 115
Counterexample, 133, 173
Counting number, vi
Crossing, 122
Cube, 127
Cube root, 50, 79
Cube root of two, 49, 74